Z
6736
.S79
1988

Stark, Marilyn
McAnally.

Mining and mineral
industries

$38.50

**Oryx Sourcebook Series
in Business and Management**

Mining and Mineral Industries

An Information Sourcebook

Oryx Sourcebook Series in Business and Management

Paul Wasserman, Series Editor

Oryx Sourcebook Series
in Business and Management

Mining
and Mineral
Industries
An Information
Sourcebook

by Marilyn McAnally Stark
Assistant Director, Information Services, Arthur Lakes Library,
Colorado School of Mines

Phoenix • New York
ORYX PRESS
1988

The rare Arabian Oryx is believed to have inspired the myth of the unicorn. This desert antelope became virtually extinct in the early 1960s. At that time several groups of international conservationists arranged to have 9 animals sent to the Phoenix Zoo to be the nucleus of a captive breeding herd. Today the Oryx population is over 400 and herds have been returned to reserves in Israel, Jordan, and Oman.

Copyright © 1988 by
The Oryx Press
2214 North Central at Encanto
Phoenix, Arizona 85004-1483
Published simultaneously in Canada

Printed and Bound in the United States of America

∞ The paper used in this publication meets the minimum requirements of American National Standard for Information Science—Permanence of Paper for Printed Library Materials, ANSI Z39.48, 1984.

Library of Congress Cataloging-in-Publication Data

Stark, Marilyn McAnally.
    Mining and mineral industries : an information sourcebook / by Marilyn McAnally Stark.
    p. cm. — (Oryx sourcebook series in business and management; no. 10)
    Includes index.
    ISBN 0-89774-295-8
    1. Mineral industries—Bibliography. 2. Mines and mineral resources—Bibliography. I. Title. II. Series.
Z6736.S79 1988
[HD9506.A2]
016.332—dc19   87-23189

Contents

Introduction

Mining and mineral industries information is broadly based, from exploration for ore deposits to processing and marketing the ores. In recent years mining has become continuously more international, and economic and political impacts of mineral industries are worldwide. Informed business decisions require both technical and management data, so both types of information are covered in this bibliography of information sources.

The literature selected varies from introductory to extremely technical and scholarly, with descriptive annotations that include level of reader's expertise assumed. English-language publications predominate, although some foreign language items, such as polyglot dictionaries and conference proceedings with papers in various languages have been included. General business sources are not included, unless a significant portion relates directly to the mineral industries. Metallurgical materials are also not included, except for a few which treat ore preparation.

Most materials are relatively recent, from the 1970s or 1980s, to reflect the changing nature of the mineral industries, companies involved, and new technology. Many conferences involving current analysis and description of mining practices have been selected. The new mining trends of computer-assisted planning and management, operations research, and robotics often occur in the conference literature.

Resource surveys of particular commodities or large areas are included, but not state-by-state or district surveys. Gold receives some emphasis, which reflects current economic conditions and interest in the mineral industries. Other surveys compare energy industries, such as coal and uranium, with petroleum. Databases and journals are a major source of mining and mineral information, and as such, receive extensive coverage.

This sourcebook is arranged according to type of reference materials, such as dictionaries, indexes, and databases. General texts and treatises are subdivided according to subject emphasis. Government documents, as an integral part of mining and mineral industries literature, are integrated within the appropriate categories along with other literature. Author, title, and subject indexes keyed to citation number appear at the end.

Mining
and Mineral
Industries

Core Library Collection

This core collection in mining and mineral industries contains those sources necessary for quick access to information, an overview of the industry, and bases for further research. These selected sources are relevant to the entire industry or selected subsets, and they are generally relatively new or serial. Journals are listed at the end of the section.

GENERAL

1. American Bureau of Metal Statistics. *Non-Ferrous Metal Data.* Secaucus, NJ, 1920– . (Annual)
> Production, supply and use of copper, lead, zinc and other nonferrous metals are published yearly in July. Statistical data from the previous year provide current information; time series can easily be constructed from past volumes in this excellent series.

2. American Metal Market. *Metal Statistics.* New York: Fairchild, 1904– . (Annual)
> Production, prices and consumption of metals are summarized briefly, then statistics, detailed breakdown of types of metals, and where to buy metals and metal products. Detailed index makes this comprehensive yearbook easy to use, and the comprehensive up-to-date coverage makes it important.

3. Carmichael, Robert S. *Handbook of Physical Properties of Rocks.* Boca Raton, FL: CRC Press, 1982. 3 vols.
> Evaluated and selected set of data on physical properties of rocks. Composition, magnetic, engineering, and mechanical properties of rocks are presented in tabular form, with graphs illustrating some values. Interdisciplinary approach, so is useful for geotechnical engineering, mining and allied fields. A reliable compendium of data, indexed by type of rock and physical property.

4. Chamberlain, Peter G., and Pojar, Michael G. *Gold and Silver Leaching Practices in the United States.* U.S. Bureau of Mines Information Circular 8969, Washington, DC: Government Printing Office, 1984. 47 p.

Leaching techniques and problems, mostly heap leaching, have become more common in the U.S. recently. State-of-the-art information for an important type of mining.

5. Coal Age. *Directory of Mine Supply Houses, Distributors and Sales Agents.* New York: McGraw-Hill, 1982. 256 p.

Mining products, manufacturers and field served are organized by state, then alphabetically. The index by company then provides access to data in this directory of U.S. firms.

6. Consolidated Gold Fields PLC. *Gold 1986.* London, 1986. 72 p.

Excellent overview of the gold industry worldwide, with gold supply and demand summarized annually in a readable, brief report. In 1987 gold is the premier exploration and development target. Prices and investment outlook are surveyed in this essential part of today's mineral industry collection.

7. Conveyor Equipment Manufacturers Association. *Belt Conveyors for Bulk Materials.* 2d ed. Boston: CBI Publishing, 1979. 346 p.

The basic source of information on belt conveyors, which are heavily used in mining. The major manufacturers association surveys economics, design, and specific components of belt conveyors. Geared toward both management and engineers, with extensive diagrams, formulas and tables of data.

8. Coope, Brian, ed. *Industrial Minerals Directory.* Surrey, England: Metal Bulletin Books, Ltd., 1977– .

Updated irregularly, new editions in 1982 and 1986, a valuable world directory of nonmetallic mineral producers, arranged by country. All nonmetallic, nonfuel minerals, including selected construction materials, are covered. Location, products, annual capacity and consuming industries are listed.

9. Crickmer, Douglas F., and Zegeer, David A., eds. *Elements of Practical Coal Mining.* 2d ed. New York: Society of Mining Engineers of AIME (The American Institute of Mining, Metallurgical, and Petroleum Engineers), 1981. 847 p.

An overview of coal mining fundamentals for those just entering the coal industry. Simplified diagrams, many photographs, little mathematics, and easy equations can help the inexperienced learn coal mining fundamentals.

10. David, Michel. *Geostatistical Ore Reserve Estimation.* Developments in Mathematics 2. Amsterdam, The Netherlands; Oxford, England; New York: Elsevier, 1977. 364 p.

Technical text for mining engineers and geologists starts with basic statistical concepts, then applies statistics to ore reserve problems. Grade-tonnage curves and orebody modeling are explained. Bibliography contains easily obtainable case studies noted as such, plus other references, and the index is detailed.

11. Deans, Alan, ed. *Alexander and Hattersley's Australian Mining, Minerals and Oil Directory.* 4th ed. Sydney, Australia: The Law Book Company Limited, 1984. 561 p.

Good overview of Australian mining activity through summary and projections. Source of considerable information on each Australian mining company with a short history plus a summary of activities. Also includes glossary of mining and oil terms.

12. E. I. DuPont de Nemours & Co. *Blasters' Handbook; A Manual Describing Explosives and Practical Methods of Use.* Wilmington, DE: DuPont, 1959– . (Annual)

Practical guide to using explosives in mining and quarrying, detailing preferred and proven practice and precautions for safety. Periodically updated and current guide to a major portion of much mining practice.

13. *E&MJ International Directory of Mining.* New York: McGraw-Hill, 1968– . (Annual)

Essential directory for the mineral industry, the 1986 edition has 607 pages of mining company data from around the world. Headquarters organizations are listed alphabetically, then mine plant units, including mines, smelters and refineries are listed by country. Financial services also included, along with industry surveys and new plant expansion. Although data are not always completely current, this is still the standard required directory for world mining and milling. Title has varied.

14. Earll, F. N., et al. *Handbook for Small Mining Enterprises.* Montana Bureau of Mines and Geology Bulletin 99. Butte, MT: Montana Bureau of Mines and Geology, 1976. 218 p.

Comprehensive and practical handbook for technical and nontechnical readers is written clearly. From geology through mining methods, reclamation, accounting and financing, with many sample forms and examples of practical applications.

15. *Environment Reporter: Mining.* Washington, DC: Bureau of National Affairs, 1978– . (Weekly)

Text of actual surface mining laws and regulations, federal and state, is reproduced in loose-leaf format for binders, beginning with the Surface Mining Control and Reclamation Act of 1977. Weekly updates note the date of legal and regulatory changes, and the new text incorporates the changes in the appropriate places. The amended section is thus complete and current. In changing regulatory climate, dependable updates of at least a portion of mining requirements are very helpful.

16. Financial Times. *Mining 1986: Mining International Year Book 1986.* Essex, England: Longman. 546 p.

The comprehensiveness of coverage and amount of information of mining companies makes this an important source. The short descriptions of properties and operations for the last three years are organized alphabetically under company name in this annual publication. Indexes by product and company, with cross reference from subsidiary to parent company, increase access points.

17. Gardiner, C. D., ed. *Canadian Mines Handbook.* Toronto, ON: The Northern Miner, 1931– . (Annual)

Brief entry by Canadian company includes financial data, production, reserves, properties, and company stock information. Entries for previous names and defunct companies refer to current entries, give date of extinction, or state that address is unknown. Mining share prices on Canadian stock exchanges and maps of mining districts complete this valuable guide to Canadian mines and mining companies.

18. Given, Ivan A., ed. *SME Mining Engineering Handbook.* New York: Society of Mining Engineers of The American Institute of Mining, Metallurgical and Petroleum Engineers, Inc., 1973. 2 vols.

The essential handbook for surface and underground mining covers practical, economical methods of mining, mineral processing and mill design. Equipment types are recommended and general costs discussed. Technical but readable, and necessary in any mining collection for managers and engineers.

19. Guilbert, John M., and Park, Charles F., Jr. *The Geology of Ore Deposits.* New York: W. H. Freeman, 1986. 985 p.

Newest version of the standard, *Ore Deposits.* Very useful for deposit descriptions, classed by type, which often cover specific mines or districts. The extensive bibliographies and detailed index just add to the overall value of this essential volume.

20. Harfax. *Guide to the Energy Industries.* Cambridge, MA: Ballinger Publishing Co., 1983. 328 p.

Nontechnical guide to financial and marketing information on coal and nuclear energy, among others. Annotated bibliography offers useful access to further information.

21. *Hart's Rocky Mountain Mining Directory 1985/86.* Denver, CO: Hart Publications, 1986. 368 p.

Mining companies, key personnel, type of business and ownership in Arizona, Colorado, Idaho, Montana, Nebraska, Nevada, New Mexico, North Dakota, South Dakota, Utah and Wyoming. Very complete listing of mining companies, joint ventures, and partnerships for any facility that operates in, or has responsibility for, areas in the Rocky Mountains. Metals, nonmetals, coal and oil shale are all covered, indexed by mine or mill operator. Unfortunately, no update is planned, at least in 1987.

22. Hoskins, J. R. *Mineral Industry Costs.* Spokane, WA: Northwest Mining Association, 1982. 248 p.

Detailed costs, including operating costs, for labor and materials from exploration through electric power, are included in this publication. Although geared to the northwest states, the data and general conclusions are applicable elsewhere.

23. Hustrulid, William A., ed. *Underground Mining Methods Handbook.* New York: Society of Mining Engineers of AIME, 1982. 1,754 p.
The new standard handbook on underground mining includes planning, equipment, ground control, and ventilation. Industry, government and academic experts contributed sections, which include many practical case studies. Profusely illustrated with diagrams. The twenty-seven-page index is very complete in this essential work.

24. *Keystone Coal Industry Manual.* New York: McGraw-Hill, 1918– .
The essential directory for the coal mining industry. Mines are listed by state, with company officers or mine superintendent, and number of employees. Coal seams are described by state, transportation is covered, and consumers and associations listed. Large and small operations are both listed, and an excellent, detailed index provides access.

25. LeFond, Stanley J., et al. *Industrial Minerals and Rocks.* 5th ed. New York: Society of Mining Engineers of the American Institute of Mining, Metallurgical and Petroleum Engineers, Inc., 1983. 1,446 p. 2 vols.
The standard reference for industrial minerals. One section grouped by uses, such as ceramic raw materials; another section by commodity, including flowsheets, production, mining, and beneficiation. An excellent index of properties, commodities, and locations speeds access.

26. Macdonald, A. James, ed. *Gold '86; An International Symposium on the Geology of Ore Deposits.* Proceedings Volume. Toronto, ON: Gold '86, 1986. 517 p.
Many newly discovered deposits in Canada, plus the United States, Europe, Africa, India, and Australasia. New information and reinterpretations of older deposits pinpoint the current focus on gold.

27. Maley, Terry S. *Mineral Title Examination.* Boise, ID: Mineral Land Publications, 1984. 396 p.
A readable, broad-ranging reference covers hard minerals, industrial minerals, coal, oil and gas for geologists, lawyers and government personnel. A glossary and detailed index access the federal actions affecting mineral status, public lands disposals, and definitions pertaining to mineral titles.

28. Maley, Terry S. *Mining Law from Location to Patent.* Boise, ID: Mineral Land Publications, 1985. 597 p.
All aspects of the General Mining Law of 1872, which is still a major part of mining law, are interpreted in plain language for a wide audience.

29. Martin, James W., et al. *Surface Mining Equipment.* 1st ed. Golden, CO: Martin Consultants, 1982. 455 p.
An overview of equipment selection considerations based on equipment design as well as mine plan. The listing of machinery by manufacturer includes specifications, listed to facilitate comparisons and eventual selection or to provide leads to companies to contact for more information.

30. *Metal Bulletin Handbook.* Surrey, England: Metal Bulletin Books, 1972– . 2 vols. (Annual)
Volume I contains prices in tabular form, with sources referenced alphabetically within nonferrous, then ferrous. Volume II covers consumption, imports and exports, plus international metals associations. Although it lacks an index, the data are easy to find nonetheless, and Metal Bulletin Books are timely and accurate.

31. Mining Journal. *Mining Annual Review.* London, 1935– . (Annual)
This authoritative review of mining activity is necessary for any mining reference collection. Industry experts summarize the year's action in mineral groups such as precious metals, and technical progress reports cover exploration, mining and metallurgy. The discussion by country includes political realities. *Mining Journal* coverage is uniformly excellent, and the yearly review is current and authoritative. The *Mining Annual Review* is included in a subscription to *Mining Journal.*

32. Mohide, Thomas Patrick. *Silver.* Ontario Ministry of Natural Resources, Mineral Policy Background Paper No. 20. Toronto, ON: Ministry of Natural Resources, 1985. 405 p.
Comprehensive nontechnical summary of world silver industry, major mines. Output is worldwide in scope, while related to Ontario mining. The overall view of silver in one volume makes it a useful addition to mineral industry literature, and the glossary, bibliography and index aid in access.

33. National Coal Association. *Coal Industry Employment/Production.* Washington, DC, 1982?– . (Annual)
U.S. Mine Safety and Health Administration statistics were computer-produced for individual mines producing 100,000 tons or more of coal in the given year. The detailed statistics cover number of employees, production per work-hour, total production and injuries. In addition, the aggregate statistics include all coal mines with any production in the year. Data are indexed by state, then company. The statistics give an effective annual picture of the coal industry.

34. National Coal Association. *Facts about Coal.* Washington, DC, 1987. 80 p. (Annual)
This brief pamphlet summarizes coal information from U.S. government statistics in a nontechnical manner for the general public and those unfamiliar with the industry. Nevertheless, a useful overview of the U.S. coal mining industry.

35. Nelson, Don, ed. *Mines and Mining Equipment and Service Companies Worldwide.* 2d ed. London: Methuen (Spon), 1985. 681 p.
Over 3,500 entries describe principal operating mining companies plus equipment and service companies. Geographical and company index.

36. Peng, Syd S. *Coal Mine Ground Control.* New York: Wiley, 1978. 450 p.
The standard handbook on ground control assumes a strength of materials knowledge to interpret roof control, rock mechanics, and equipment recommendations from premining to subsidence instrumentation. Many

diagrams and photographs illustrate the text, and a comprehensive index and references complete access.

37. Posner, Mitchell J. and Goldberg, Philip. *The Strategic Metals Investment Handbook.* New York: Holt, Rinehart and Winston, 1983. 372 p.

Written for the layman investor, the handbook explains strategic metals, or minor metals, such as rare-earth metals, cobalt or tungsten, as by-products of base metals. Investments in the commodities themselves, including storage and insurance, or in stocks of mining companies, are recommended, and a list of Minor Metals Traders' Association members is included. Useful for the business side of the minerals industry.

38. Pruitt, Robert G., Jr., ed. *Digest of Mining Claim Laws.* Boulder, CO: Rocky Mountain Mineral Law Foundation, 1977. 176 p.

For persons interested in evaluating claims, this focuses on location and maintenance of claims. Sections of laws are reproduced, plus interpretation of meaning, and comments under each state highlight different and unusual requirements.

39. Robbins, Peter. *Guide to NonFerrous Metals and Their Markets.* 3d ed. London: Kogan Page; New York: Nichols Publishing, 1982. 183 p.

Uses, reserves, locations, marketing and pricing of metals worldwide are summarized. The role of supply and demand in pricing is covered in an understandable manner.

40. Rocky Mountain Mineral Law Foundation. *The American Law of Mining.* 2d ed. New York: Matthew Bender, 1984– . 6 vols.

Loose-leaf service covers all aspects of mining law, from exploration through operation to closure. Updated periodically, so coverage kept current. Necessary for compliance planning concerning mines in this U.S. regulatory situation. Cumulative index published separately.

41. Schumacher, Otto, ed. *Mining Cost Service; A Subscription Cost Data Update Service.* Spokane, WA: Western Mine Engineering, 1983– .

Mining costs in the fourteen western states plus Wisconsin and Minnesota are updated several times a year by sections such as commodities, equipment, labor, transportation and taxes. Because costs change so rapidly, the updates provide current, accurate data for feasibility studies. The simple format of general cost data is a very useful and continuing resource.

42. Sloan, Douglas A. *Mine Management.* London, New York: Chapman and Hall, 1983. 495 p.

A practical, readable book focused on how to manage mines, from costing through maintenance, materials and wages. It emphasizes necessary information systems to allow effective management and principles to be applied to a specific industry.

43. Stack, Barbara. *Handbook of Mining and Tunneling Machinery.* Chichester, England: Wiley, 1982. 742 p.

This handbook is valuable for its history of mining machinery, based on the machines whose manufacturer made major contributions to that type. Patents are translated into plain language, and the development of mining machinery is placed in worldwide perspective.

44. Staley, William Wesley. *Introduction to Mine Surveying.* 2d ed. Stanford, CA: Stanford University Press, 1964. 303 p.

Mine surveying, mostly underground, is the focus, and specialized equipment and procedures are described. From instruments, their adjustment and use, through stope and tunnel surveys. Open-pit surveying is also covered.

45. Stout, Koehler S. *Mining Methods & Equipment.* New York: McGraw-Hill, 1980. 218 p.

Basic explanatory materials, with many diagrams of mining methods and photographs of mining machinery. This nontechnical and easy-to-understand survey covers surface and underground mining, from exploration to reclamation.

46. Thomas, Paul R. and Boyle, Edward H., Jr. *Gold Availability— World; A Minerals Availability Appraisal.* U.S. Bureau of Mines Information Circular 9070. Washington, DC: Government Printing Office, 1986. 87 p.

Emphasis on gold exploration and mining makes this brief study a necessary part of mining reference collections. History, data and statistics on worldwide gold production, costs, producers, and projections for gold production are all included in this government publication.

47. Thrush, Paul W. and the Staff of the Bureau of Mines, eds. *A Dictionary of Mining, Minerals and Related Terms.* Washington, DC: Government Printing Office, 1968. 1,269 p.

This outstanding, very complete dictionary has been reissued, because it is still the basic English-language dictionary on mining and mineral industries. Concise definitions are still very useful, even though the dictionary is nearly twenty years old. A must for every collection.

48. Touloukian, Y. S.; Judd, W. R.; and Roy, R. F., eds. *Physical Properties of Rocks and Minerals.* McGraw-Hill/Cindas Data Series on Material Properties. New York: McGraw-Hill, 1981. Vol. II-2. 548 p.

Characteristics, porosity, permeability, strength, deformability, magnetic properties of rocks and minerals are described and illustrated with statistical tables. Many references lead to further information.

49. The Uranium Institute. *The Uranium Equation: Balance of Supply and Demand 1980–95.* London: Mining Journal Books, 1981. 57 p.

An earlier edition was revised because supply and demand were influenced by major changes, less global economic growth and anti-nuclear politics.

50. U.S. Bureau of Mines. *List of Bureau of Mines Publications and Articles, with Subject and Author Index.* Washington, DC: Government Printing Office, 1961– . (Annual)

Extensive index to Bureau of Mines publications is essential for access to mining publications. *Report of Investigations, Information Circulars, Open-File Reports,* and nongovernment publications by bureau authors are covered. These supplements to *List of Publications Issued by the Bureau of Mines* are invaluable sources for mining and processing information, and the cumulative indexes every five years help make access more convenient. Abstracts are included for most entries.

51. U.S. Bureau of Mines. *List of Publications Issued by the Bureau of Mines from July 1, 1910 to January 1, 1960, with Subject and Author Index.* Washington, DC: Government Printing Office, 1960. 826 p.

Brief annotations describe *Report of Investigations, Information Circular, Minerals Yearbook,* and other Bureau publications. Extensive index uses terms specific to mining, the mineral industries and mineral processing.

52. U.S. Bureau of Mines. *Mineral Facts and Problems,* 1985 ed. Washington, DC: Government Printing Office, 1985. 956 p.

Begun in 1953, this essential reference is published every five years, and preprints of chapters are available. Commodity chapters include summary, reserves, industry, supply, demand, and economic factors, with a bibliography for each of the metals and nonmetals. The only disadvantage is the information date; the 1985 edition includes 1982 data as the most recent, and 1983 as preliminary.

53. U.S. Bureau of Mines. *Mineral Position of the United States: The Past Fifteen Years.* The 1985 Annual Report of the Secretary of the Interior under the Mining and Minerals Policy Act of 1970. Pittsburgh, PA, 1986. 45 p.

Performance of nonfuel mineral industry during 1985 is reported, in a brief format, with demand and competition and the U.S. responses.

54. U.S. Bureau of Mines. *Minerals Yearbook.* Vol. I—*Metals and Minerals*; Vol. II—*Area Reports, Domestic*; Vol. III—*Area Reports: International.* Washington, DC: Government Printing Office, 1932– . (Annual)

To understand the status of worldwide mineral industries, these annual volumes are essential, even though the data are generally two or more years old. Lengthy articles and data summarize production and events that affect that production. *Volume I—Metals and Minerals* covers quantity and value of production, including trade and prices. *Volume II—Area Reports, Domestic* covers industry by state, including quantity, value, and legislation affecting production. *Volume III—International* summarizes history, production amounts, trade, principal trading partners and government policies. The entire series is published annually, and preprints of specific sections are available. Inexpensive and necessary detailed surveys.

55. U.S. Bureau of Mines. *New Publications—Bureau of Mines.* Washington, DC: Government Printing Office, 1984– . (Monthly)

Monthly listing of Bureau of Mines publications and articles by mines staff is a brief pamphlet of new items. Abstracts are included for most items. Cumulated annually into *List of Bureau of Mines Publications and Articles, with Subject and Author Index* and every five years in another cumulation. All the Bureau of Mines indexes are necessary for any basic mining collection.

56. U.S. Forest Service. Intermountain Region, Ogden, Utah. Surface Environment and Mining. *Anatomy of a Mine from Prospect to Production.* U.S. Department of Agriculture General Technical Report INT-35. Ogden, UT, 1977. 69 p.

The development of a mine is traced in plain language with basic, understandable illustrations. Intended for land managers, planners, mining industry personnel, and environmentalists. Valuable to explain clearly the complexities of bringing a mine to the point of actual production.

57. Vick, Steven G. *Planning, Design, and Analysis of Tailings Dams.* Wiley Series in Geotechnical Engineering. New York: Wiley, 1983. 369 p.

A practical orientation on tailings dams, an integral part of many mining projects, from design to reclamation. Although precise and technical, explanations are straightforward and comprehensible for geotechnical, mining and metallurgical engineers, plus regulatory personnel. A sixteen-page reference list and comprehensive index add easy access and further information if needed.

58. Wanless, R. M. *Finance for Mine Management.* New York: Chapman and Hall, 1983. 208 p.

Basic information, directed to managers without financial background, considers items to get a typical mine ready for production. Risk evaluation and financial accounting included, and an extensive index provided in this basic study of an integral part of mineral production.

59. *Western Mining Directory.* Denver, CO: Howell Publishing Company, 1978– . (Annual)

Current directory of active mines, mining companies, consultants, contractors, equipment companies involved in exploration, drilling, production and milling in the fourteen western states of the U.S. Complete and detailed listing of companies active in the west.

60. Williams, Roy E. *Waste Production and Disposal in Mining, Milling, and Metallurgical Industries.* A World Mining Book. San Francisco, CA: Miller-Freeman, 1975. 489 p.

Mining and milling wastes, treatment methods and reclamation are characterized with a water pollution emphasis, based on requirements of Federal Water Pollution Control Amendments of 1972 (Public Law 92–500). Effluent limitations guidelines for point source discharges from mineral resource-based industries. Cost data and flowsheets are included, along with a detailed index, in this standard reference.

61. World Coal Study. *Coal: Bridge to the Future.* Cambridge, MA: Ballinger, 1980. 247 p.

This often-quoted World Coal Study, or WOCOL, evaluates coal's importance in the energy market compared to oil, gas and nuclear power. Even with the uncertainty of the world's energy system, the study is optimistic about energy. Transportation, market and environment in the sixteen countries which use 75 percent of world energy are summarized. Useful for its comparative assessments.

62. Wyllie, R. J. M., and Argall, George O., Jr. *World Mining Glossary of Mining, Processing, and Geological Terms.* Rev. ed. San Francisco, CA: Miller-Freeman Publications, 1975. 432 p.

This polyglot mining glossary lists English words and phrases alphabetically along with equivalents in Swedish, German, French and Spanish. Words in the non-English languages are then indexed to the English words. The basic multilanguage mining glossary.

63. Zack, Gene J. *Pit & Quarry Directory of the U.S. Nonmetallic Mining Industries.* Chicago: Pit and Quarry Publications, 1982. 480 p.

Directory of sand and gravel, aggregate, and stone producers in the U.S., arranged by state and company. Name of mine site, product mined, and nearest town given for each entry. New editions produced irregularly, the previous in 1970, are very useful for basic company information needed in a mining collection.

JOURNALS

64. *Coal Age.* New York: McGraw-Hill, 1911– . (Monthly)

Fairly lengthy news on the coal industry is featured, plus technical articles on coal-mine equipment, design, operation and regulation. Good coal calendar.

65. *Coal Mining.* Chicago: Maclean Hunter, 1964– . (Monthly)

U.S. and Canadian coal industry and market information is geared toward management and technical personnel with general interest articles. Necessary journal to keep up with coal mining. Previous title was *Coal Mining & Processing.*

66. *E&MJ Engineering and Mining Journal.* New York: McGraw-Hill, 1866– . (Monthly)

One of the foremost journals in mining, *E&MJ* covers worldwide, broad-based news, and feature articles often include mine data, costs and flowsheets. Written for the practicing engineer and manager, the information is timely and practical. March issue contains annual survey and outlook for mineral commodities.

67. *Metals Week.* New York, McGraw-Hill, 1930– . (Weekly)
Daily, weekly, monthly and annual prices of metals listed in this newsletter are used as authoritative data in many other publications. This authoritative publication in the minerals industry also offers economics-oriented brief news which provide quick information on supply and demand in the metals industry.

68. *Mining Engineering.* Littleton, CO: Society of Mining Engineers of AIME, 1949– . (Monthly)
The major U.S. mining society publishes this standard mining journal, which emphasizes mineral exploration and production of coal and metals, plus personnel news and industry outlook, both in the U.S. and worldwide. Most entries are written by working mining engineers and reflect current mining practice and outlook.

69. *Mining Journal.* London: Mining Journal, 1835– . (Weekly)
The best current coverage of the worldwide mineral industries. One lead article which analyzes commodity or country mining status is joined by other brief news about a column long, all of which provide the most complete review of developments in the mineral industry. Gold is reviewed quarterly in this newsletter.

70. *Mining Magazine.* London: Mining Journal, 1909– . (Monthly)
This magazine, which comes as part of a subscription to *Mining Journal,* presents more detailed information on worldwide mining, with practical information on particular mines, such as technical data, specifications and costs.

71. *Northern Miner.* Toronto, ON: Northern Miner Press, 1915– . (Weekly)
Newspaper covers Canadian mining and mineral industries, plus activities of Canadian companies throughout the world. Exploration is widely reported, as well as operating mines. Good coverage of Latin American mining through connection with Canadian companies. Necessary journal to provide well-rounded coverage of mining industries.

72. *Transactions of the Society of Mining Engineers of AIME.* Littleton, CO: SME, 1871– . (Annual)
Yearly hardbound volume of information on mineral exploration, mining and processing. Some papers were previously printed in *Mining Engineering* and *Minerals & Metallurgical Processing,* and some include new material to expand the technical literature on mineral production.

Dictionaries and Encyclopedias

Dictionaries and encyclopedias for mineral industries and mining are limited in number and are older than the rest of document types. Nevertheless, some subject-specific areas of mining are covered, and translating dictionaries are available.

73. Adamov, N. *Dictionnaire du Petrole et des Mines: Francais-Anglais, Anglais-Francais.* Paris: Paris-Monde, 1963. 660 p.
Alphabetical arrangement of mineral industries and petroleum, first in French, then in English.

74. Gilpin, Alan. *Dictionary of Fuel Technology.* New York: Philosophical Library, 1969. 279 p.
All types of fuel, including coal and nuclear as well as oil and gas, are defined. Commercial names and organizations are included in addition to the scientific terms.

75. National Coal Board. *Terminology of Remote Control and Automation in Mining.* London, 1979. Unpaged.
Programming, engineering, and mathematical terms are arranged alphabetically, with equivalent French and German terms. Index in French and German.

76. Picot, P., and Johan, Z. *Atlas of Ore Minerals.* Translated by J. Guilloux. Amsterdam, The Netherlands: Elsevier, 1982. 458 p.
Ore minerals, their characteristic reflectance, texture and occurrence, are defined, often accompanied by color photographs. Specimens from Bureau de Recherches Geologiques et Minieres, France.

77. Schmidt, Richard A. *Coal in America: An Encyclopedia of Reserves, Production and Use.* New York: McGraw-Hill, 1979. 458 p.
Coal properties and reserves, production and use allow evaluation of current trends and issues in coal development.

78. Schwicker, Angelo Cagnacci. *International Dictionary of Metallurgy—Mineralogy—Geology.* Milan, Italy: Technoprint International, 1968. 1530 p.

English-French-German-Italian polyglot dictionary on the mining and oil industries. Words and phrases in English are translated into the other languages. Indexes are translated from the other languages into English.

79. Somerville, S. H. and Paul, M. A. *Dictionary of Geotechnics.* London: Butterworth, 1983. 283 p.

Engineering geology dictionary which covers soil and rock mechanics and engineering. The brief definitions have an English orientation and worldwide coverage.

80. Sutulov, Alexander, ed. *International Molybdenum Encyclopedia 1778-1978.* Santiago, Chile: Sutulov, 1978. 3 vols.

Alphabetical entries of varying length, from a brief paragraph to eight or ten pages, in this comprehensive treatise on molybdenum. The three volumes are I—Resources and Production; II— Metallurgy and Processing; III—Products, Uses and Trade. Companies, ore deposits, mines, processes, and production are all included.

81. Todd, Arthur H. J. *Lexicon of Terms Relating to the Assessment and Classification of Coal Resources.* London: Graham & Trotman, 1982. 136 p.

Precise terms from authoritative documents include references back to source in each definition. English orientation.

82. Tomkeieff, S. I. *Dictionary of Petrology.* Edited by E. K. Walton et al. Chichester, England; New York: Wiley, 1983. 680 p.

Dictionary with brief definitions, code for major category or type of rock, and reference source for each definition. Synoptic tables regroup terms into lists pertaining to differenct subjects in petrology. Extensive references to European literature in this scholarly dictionary.

Handbooks and Manuals

Handbooks and manuals vary from "how-to-do-it" instructions to reviews of state-of-the-art in an aspect of mining. Most are technical and recommend procedures and equipment and are designed to be consulted to solve specific problems or apply new technologies.

83. Bauer, Eric R. *Ground Control Instrumentation; A Manual for the Mining Industry.* U.S. Bureau of Mines Information Circular 9053, Washington, DC: Government Printing Office, 1985. 68 p.
Manual on technology for measuring ground movement. Convergence, roof movement, stress and support are discussed and instrumentation selection procedures presented. Case studies and a list of instrument suppliers included in this basic handbook.

84. Canadian Mining Journal. *Reference Manual & Buyers Guide, 1891– .* Don Mills, Canada: Southam Business Publications. (Annual)
Technical mining and mineral processing methods and costs are examined for exploration costs and budget for underground and open pit mines plus mills. An index of equipment is included.

85. Chironis, Nicholas, ed. *Coal Age Operating Handbook of Coal Surface Mining and Reclamation.* Coal Age Library of Operating Handbooks, vol. 2. New York: McGraw-Hill, 1978. 442 p.
Coal Age articles from 1974–78 discuss outstanding surface mines, new concepts in surface mining, and research and development. Bulldozers, blasting equipment, trucks and other equipment are covered, and a subject index is provided.

86. Chironis, Nicholas, ed. *Coal Age Operating Handbook of Underground Mining.* Coal Age Library of Operating Handbooks, vol. 1. New York: McGraw-Hill, 1977. 410 p.
Coal Age articles on underground mining from 1973–77 are reprinted in this handbook which focuses on increased production and reduced accidents. A subject index is included.

87. Chironis, Nicholas, ed. *Coal Age Second Operating Handbook of Underground Mining.* Coal Age Library of Operating Handbooks, vol. 4. New York: McGraw-Hill, 1980.
Coal Age editors wrote the volume, with articles and data contributed by mining experts. Techniques, longwall mining, haulage, roof control, ven-

tilation, transportation and management are covered, and entries are indexed.

88. Clement, George K. *Capital and Operating Cost Estimating System Manual for Mining and Beneficiation of Metallic and Nonmetallic Minerals Except Fossil Fuels in the United States and Canada.* U.S. Bureau of Mines Special Publication 4–81. Washington, DC: Government Printing Office, 1981. 149 p.

Minerals Availability System (MAS) is explained. Physical and commercial availability of miner resources are defined, and cost curves and equations for calculations are presented.

89. *Coal Age Equipment Guide.* New York: McGraw-Hill, 1982. 230 p.

Articles from *Coal Age* compare various types of machinery, including specifications, facts and figures. *Coal Age* editors wrote most of the guide, but dates for original publication are not included. Indexes are by author, company and subject.

90. Crowson, Phillip. *Minerals Handbook 1986–1987.* Houston, TX: Gulf Publishing Company, 1986. 331 p.

U.S. Bureau of Mines, *Metal Bulletin,* and other statistical sources first, provide broad coverage then by commodity. Statistics are from 1984–85. Individual statistical tables are sometimes not referenced to sources.

91. Dick, Richard; Fletcher, Larry R.; and D'Andrea, Dennis V. *Explosives and Blasting Procedure Manual.* U.S. Bureau of Mines Information Circular 8925. Washington, DC: Government Printing Office, 1982. 105 p.

Practical manual on state-of-the-art technology in explosives and blasting is compiled from U.S. Bureau of Mines research, mining industry, and explosives companies.

92. E. D'Appolonia Consulting Engineers, Inc. *Engineering and Design Manual: Coal Refuse Disposal Facilities.* Washington, DC: U.S. Mining Enforcement and Safety Administration, 1975. 862 p.

Comprehensive study of design of U.S. coal refuse facilities. Characteristics of coal refuse, geotechnical studies recommended, operations and monitoring, instrumentation and environmental impacts are considered. Extensive references.

93. Environmental Policy Institute and Center for Law and Social Policy. *The Strip Mine Handbook.* Washington, DC: EPI, 1978. 107 p.

Geared toward citizens who want to monitor and report coal mining violations, this handbook points out possible violations and links them to the appropriate section of the Surface Mining Control and Reclamation Act (Public Law 95–87).

94. Kulczak, Frank. *Illustrated Surface Mining Methods.* New York: McGraw-Hill, 1979. 87 p.

Arranged by geographic region and type of mining within that region, diagrams of surface-mining methods are presented for mining personnel and those outside the field. Easy to understand, clear illustrations.

95. Lester, D. *Quarrying and Rock Breaking; The Operation and Maintenance of Mobile Processing Plants.* London: Intermediate Technology Publications, Ltd., 1981. 116 p.

Very basic, easy-to-read instructions on installation, operation and maintenance of mobile processing plants. This handbook uses the cookbook approach, with emphasis on maintenance and safety of a little-discussed technology.

96. Lewis, W. H. *Underground Coal Mine Lighting Handbook (In Two Parts).* 1. Background, 2. Application. U.S. Bureau of Mines Information Circulars 9073, 9074. Washington, DC: Government Printing Office, 1986. 89, 42 p.

These complete references on underground coal mine lighting cover fundamentals of light, technical considerations, measuring, regulations, hardware, lamps and ballasts. Installation and maintenance of an appropriate and safe coal mine lighting system are well explained.

97. Martin, James W., et al. *Guideline Manual for Front-End Loader Load-and-Carry Operations.* U.S. Bureau of Mines Open-File Report 169–82, Pittsburgh, PA: U.S. Bureau of Mines, 1981. 253 p. PB-115014.

Application, selection, and performance of various front-end loaders are analyzed.

98. McAteer, J. Davitt. *Miner's Manual.* Washington, DC: Crossroads Press, 1981. 357 p.

Complete guide to job health and safety is written for the working miner, with clear, concise explanations of mining regulations under topics such as electrical hazards, roof control, and noise. Mine Safety and Health Administration (MSHA) regulations are interpreted and examples of MSHA Fatal Accident Investigation Report included.

99. Merritt, Paul C. *Coal Age Operating Handbook of Coal Preparation.* Coal Age Library of Operating Handbooks, vol. 3. New York: McGraw-Hill, 1978. 311 p.

Reprints from *Coal Age* are joined by contributed articles. Equipment, processes, design factors and evaluation criteria are subjects. A portfolio of flowsheets and the American Society for Testing and Materials (ASTM) test standards for coal are included. A subject index provides access.

100. Meyers, Robert A. *Coal Handbook.* New York: Marcell Dekker, 1981. 854 p.

Overall survey of coal science and technology includes coal characteristics, sampling and analysis, surface and underground coal mining, and coal cleaning. Both a glossary and index are included.

101. Mining and Reclamation Council of America. *Permit Application Handbook.* Washington, DC, 1981. 447 p. DE81027149.
Cookbook approach for small mine operators to follow to comply with Federal Surface Mining Control and Reclamation Act of 1977. Requirements, such as site description, operation plan and reclamation plan, are explained, and a checklist of necessary components is included.

102. Mular, A. R. *Mining and Mineral Processing Equipment Costs and Preliminary Capital Cost Estimations.* The Canadian Institute of Mining and Metallurgy, special vol. 25. Montreal, PQ: The Canadian Institute of Mining and Metallurgy, 1982. 265 p.
Text, graphs and tables for estimating cost of major equipment types in mining and mineral processing. Examples illustrate the key components of the cost, and cost indexes allow updating.

103. National Coal Board. Mining Department. *Subsidence Engineers' Handbook.* 2d rev. ed. London, 1975. 111 p.
The standard handbook for all phases of subsidence control. British focus but applicable in other parts of the world.

104. Nunenkamp, David C. *Coal Preparation Environmental Engineering Manual.* McLean, VA: J. J. Davis Associates, 1977. 727 p. EPA-600/2–76–138.
Coal preparation and its environmental impact are assessed. Coal characteristics, cleaning equipment and resulting wastes are discussed, and environmental mitigation presented.

105. Padley & Venables, Ltd. *Rock Drilling Data.* Sheffield, England, 197?. 224 p.
Drilling and blasting data and practice presented for field sales engineers. Proper practice and maintenance are discussed, and many diagrams illustrate the process. Failures are explained and reasons given, plus advice on how to avoid each type of failure.

106. Pearl, Richard M. *Handbook for Prospectors.* 5th ed. New York: McGraw-Hill, 1973. 472 p.
Practical guidelines for prospecting for different kinds of minerals and ores, with sections on basic prospecting, then particular types of ores and their recognition. Glossary and index.

107. Peele, Robert. *Mining Engineers' Handbook.* 3d ed. New York, Wiley, 1941. 2 vols.
This edition is still in print and sometimes relevant to underground mining, mostly of minerals. A little information on coal mining. Sampling, assaying and testing are covered. The *SME Mining Engineering Handbook* has updated and broadened most sections.

108. Peng, Syd S. and Chiang, H. S. *Longwall Mining.* New York: Wiley, 1984. 708 p.
This technical, practical book for training uses examples from the eastern U.S. and China on techniques of longwall mining. Strata control, coal extraction, transportation, instrumentation, and safety considerations are all presented in this comprehensive handbook.

109. *Pit & Quarry Handbook and Buyer's Guide; Equipment and Technical Reference Manual for Nonmetallic Industry.* New York: Harcourt Brace Jovanovich, 1907– . (Annual)
Equipment-focused handbook, geared toward purchase of equipment for mines and quarries. Design, dredging, pumps, and blasting are surveyed for state-of-the-art equipment. Valuable handbook because of its focus on the nonmetallic minerals industry.

110. Pojar, Michael G. *Surface and Underground Coal Mine Equipment Population, 1982.* U.S. Bureau of Mines Information Circular 9078. Washington, DC: Government Printing Office, 1983. 25 p.
Equipment predicted from baseline 1976 data, with adjustments from the half of known surface and underground mines in 1982.

111. *Robert D. Fisher Manual of Valuable & Worthless Securities.* New York, 1926–71. 15 vols.
Series on extinct or obsolete companies covers many early mining firms. Usually date of cessation of operations is given, date of charter forfeit, or date of absorption into another company. Basic for checking value of early mining stocks. Indexes by sets of volumes. Title varies, earlier volumes are *Mervyn Scudder Manual of Extinct or Obsolete Companies.*

112. Schenk, George C. K. *Handbook for State and Local Taxation of Solid Minerals.* U.S. Bureau of Mines Open-File Report 38–84. Pittsburgh, PA: U.S. Bureau of Mines, 1983. 181 p. PB 84–166602.
Economic analysis of new or changed taxes on owners and miners of mineral deposits. Income, property and severance taxes are analyzed and relative advantages calculated.

113. Shirey, Glenn A.; Colinet, Jay F.; and Kost, John A. *Dust Control Handbook for Longwall Mining Operations.* U.S. Bureau of Mines Open-File Report 34–86. Pittsburgh, PA: U.S. Bureau of Mines, 1985. 223 p.
Detailed handbook on dust control methods for longwall mining, design explanations and relationship to control procedures. Extensive reference lists.

114. Sisselman, Robert, ed. *E/MJ Operating Handbook of Mineral Underground Mining.* E/MJ Library of Operating Handbooks, vol. 3. New York: McGraw-Hill, 1978. 440 p.
Articles from the 1975–78 issues of *Engineering and Mining Journal* are cumulated in one volume for reference. Practical information sometimes illustrates practices in particular mines.

115. Skelly and Loy. *Training Manual for Miners; Follows MSHA's Guidelines.* Edited by P. Nicholas Chironis. New York: McGraw-Hill, 1980. 2 vols.
Complete study guide for training miners, with self-tests. Safety orientation is presented, reasons why some things must be done is explained, and potential problems are noted. The miner's rights are emphasized. Volume 1 covers underground mining; Volume 2, surface.

116. Stout, Koehler S. *The Profitable Small Mine: Prospecting to Operation.* New York: McGraw-Hill, 1984. 174 p.
Practical handbook showing how to prospect, sample and estimate resources, develop, operate, and maintain the small mine. Equipment and costs are covered.

117. Taggart, Arthur F. *Elements of Ore Dressing.* New York: Wiley, 1951. 595 p.
Crushing, dewatering, and metallurgy are presented and machinery types defined and illustrated. An index is included. Somewhat dated but still basic.

118. Taggart, Arthur F. *Handbook of Mineral Dressing; Ores and Industrial Minerals.* New York: Wiley, 1945. 1744 p.
Although somewhat dated, this handbook on mineral preparation, washing, drying and treatment, and dressing has not been supplanted and is still useful. Later edition of *Handbook of Ore Dressing.*

119. Thomas, R., ed. *E/MJ Operating Handbook of Mineral Processing: Concentrating, Agglomerating, Smelting, Refining, Extractive Metallurgy.* E/MJ Library of Operating Handbooks, vol. 1. New York: McGraw-Hill, 1977. 426 p.
Reprints of applicable articles from *Engineering & Mining* Journal, E/MJ cumulated in one bound volume for easy reference.

120. TRW, Inc. *Oil Shale Data Book.* Springfield, VA: National Technical Information Service, 1979. 401 p. PB 90–125636.
Handbook covers oil shale mining, processing, retorting of ore, transportation and waste disposal—the whole range of oil shale handling.

121. U.S. Bureau of Mines. *Splicing Mine Cables.* U.S. Bureau of Mines Handbook HB 1–84. Washington, DC: Government Printing Office, 1984. 92 p.
Procedures on splicing, common errors and how to avoid them. Practical, easy to understand handbook geared toward the working miner.

122. U.S. Mining Enforcement and Safety Administration. *Coal Mine Health and Safety Inspection Manual for Surface Coal Mines and Surface Work Areas of Underground Mines.* Washington, DC: MESA, 1974. 43 p.
Manual analyzes Coal Mine Health and Safety Act of 1969 and its accompanying regulations for surface coal mines. The handbook is geared to mine inspectors and coal industry personnel.

123. U.S. Mining Enforcement and Safety Administration. *Coal Mine Health and Safety Inspection Manual for Underground Mines.* Washington, DC: MESA, 1974. 121 p.
Manual analyzes compliance with Coal Mine Health and Safety Act of 1969 and its accompanying regulations for underground coal mines. The handbook should assist mine inspectors in performing safety inspections and industry personnel in knowing what to expect.

124. U.S. Mining Enforcement and Safety Administration. *Coal Mining Safety Manual No. 1.* Washington, DC: Government Printing Office, 1974. 34 p.
Nontechnical explanation of coal, its origin and mining. General discussion of safety practices in coal mining and requirements for safe mining.

125. U.S. Mining Enforcement and Safety Administration. *Coal Refuse Inspection Manual.* Washington, DC, 1976. 121 p.
This handbook presents an overview of engineering and environmental aspects of coal waste disposal. Identification of coal refuse types, critical stability requirements, instability, and how to inspect the facilities are covered.

126. White, Lane, ed. *E&MJ Second Operating Handbook of Mineral Processing.* E&MJ Library of Operating Handbooks, vol. 4. New York: McGraw-Hill, 1980. 509 p.
Articles from *Engineering and Mining Journal,* 1977–79, detail current practice in mineral processing and plant operation. Straightforward presentation of earlier articles.

Directories

Mining directories comprise several types: biographical, company and technical. Mining engineers are included in several technical directories, and indexes often include specialty access. Company directories often include mining executives as part of the entries. Technical directories list such items as computer programs and their characteristics.

BIOGRAPHICAL

127. *CIM Directory.* Montreal, PQ: Canadian Institute of Mining and Metallurgy, 1967– . (Annual)
   Canadian mining and metallurgical engineers are listed with addresses, plus association officers and a description of the organization.

128. Davis, Gordon. *Who's Who in Engineering.* 6th ed. New York: American Association of Engineering Societies, 1985. 1,900 p.
   Brief biographical information on engineers and academics in engineering. Contributions to science publications are listed, and mining engineers are among the categories covered.

129. *Modern Scientists and Engineers.* New York: McGraw-Hill, 1980. 3 vols.
   One- to two-page biographies describing outstanding achievements and reasons for entry, plus academic credentials. Some entries for geological engineers. Index is by name and specialty.

130. Pelletier, Paul A. *Prominent Scientists: An Index to Collective Biographies.* New York: Neal Schuman, 1980. 311 p.
   Index to scientists' biographies that appear in books of collective biography, mostly 1960–79. Alphabetical by author, with list by discipline, including mining.

131. Society of Mining Engineers of AIME. *Who's Who in Mineral Engineering: SME Membership Directory.* Littleton, CO: SME, 1949– . (Annual)
   Name, employment, and address of members of SME are published as the July issue of *Mining Engineering.*

132. Tinucci, Barbara A., ed. *Who's Who in Technology Today.* 5th ed. Lake Bluff, IL: J. Dick & Company, 1984. 5 vols.
Mining personnel are covered in Volume 4, with brief biographical information on education, patents, achievements, honors, and area of principal expertise. Index is Volume 5.

133. *Who's Who in Frontier Science and Technology.* New York: Marquis Who's Who, 1984–85. 846 p.
Experts in the forefront fields of technology are accorded very brief listings of accomplishments and education. Access by index under energy science and technology, coal, oil shale, nuclear.

COMPANIES

134. *Australian Mining Year Book 1986.* Chippendale, Australia: Thomson Publications, 1986. 374 p.
Annual directory of Australian mining and exploration companies and their activities during the year. Financial data disclosed. Products and suppliers indexes, plus organizations, ore buyers and government departments.

135. Nielsen, George F. *U.S. Coal Mine Production by Seam.* 3d ed. New York: McGraw-Hill, 1980. 806 p.
Coal mines listed by state and county, with type of mine and tonnage produced. Production data for each mine include proximate analyses, coal content, and production from the seam. Indexes include seam names, by U.S. Bureau of Mines code, and alphabetical by state. Overall view of production by coal seam and characteristics of that seam.

136. *1985 Coal Mine Directory.* New York: McGraw-Hill, 1985. 470 p.
Mine directory and executives sections of *Keystone Coal Industry Manual.* Intended for manufacturers and distributors.

137. *U.S. Coal Production by Company: Bituminous, Anthracite, Lignite.* New York: McGraw-Hill, 1968– . (Annual)
Principal producing U.S. coal companies of 100,000 tons or more are listed by tonnage class, and coal production is presented. Historical data are included, and the fifty largest mines are listed.

138. U.S. Energy Information Administration. *Coal Distribution Companies in the United States: 1980.* Washington, DC, 1981. 107 p. DOE/EIA-0217(80).
Alphabetical listing of coal sellers, including coal mining companies, coal brokers and coal retailers. Address and coal districts served are included.

TECHNICAL

139. Cline, Robert Lee. *Directory of Computer Models Applicable to the Coal Mining Industry.* Golden, CO: Colorado School of Mines Press, 1981. Various paging.

Computer models for surface or underground mining of coal are categorized according to intended application. Programs are described, capabilities and methodology assessed. The responsible organization and its address are listed, along with the author.

140. U.S. Bureau of Mines. *Basic Coal Research in the United States.* U.S. Bureau of Mines Information Circular 8390. Washington, DC: Government Printing Office, 1968. 56 p.

Research is categorized by type, such as analysis, dewatering, sulfur. Participating researchers are listed, subject index included.

141. U.S. Bureau of Mines. *Nonfuel Mineral Model Directory.* U.S. Bureau of Mines Information Circular 8966. Washington, DC: Government Printing Office, 1984. 40 p.

Listing of models for nonfuel minerals, responsible person and contact information.

142. U.S. Mineral Data Working Group Interagency Minerals Information Coordinating Committee. *Mineral Data Source Directory.* U.S. Bureau of Mines Information Circular 8935. Washington, DC: Government Printing Office, 1983. 376 p.

Listing of mineral data sources within the federal government and contact persons.

Bibliographies and Literature Reviews

Numerous bibliographies have been compiled on facets of mining and minerals, and representative ones are listed. When items covered are unusual, the contents are noted, otherwise books, reports, and journal articles are considered to be included.

143. Akers, David J.; McMillan, Barry G.; and Leonard, Joseph W. *Coal Minerals Bibliography.* Washington, DC: U.S. Department of Energy, 1978. 230 p. FE-2692-5.

Annotated bibliography by the Coal Research Bureau of the University of West Virginia on the inorganic minerals in coal and coal byproducts. Subject index included.

144. Alexandrov, Eugene A. *Mineral and Energy Resources of the USSR; A Selected Bibliography of Sources in English.* Falls Church, VA: American Geological Institute, 1980. 91 p.

Valuable access to sources in English on Russian mineral deposits. Most references are from *International Geology Review,* with some state-of-the-art and other materials added.

145. Bituminous Coal Research, Inc. *Reclamation of Coal-Mined Land; A Bibliography.* Monroeville, PA, 1975. 188 p.

More than 700 technical articles, conference papers, and government reports on reclamation of coal-mined land, with brief annotations. Focus on eastern U.S. but some western U.S. entries included.

146. Bloch, Carolyn C. *Federal Energy Information Sources and Data Bases.* Park Ridge, NJ: Noyes Data, 1979. 115 p.

Guide to sources of energy information in the federal government, from research centers to administrative agencies and congressional offices. Libraries are listed, along with computerized access via databases. Although some responsibilities have changed, the specific retrieval points are mostly still valid.

147. Clark, J.; Caldon, J. H.; and Curth, E. A. *Thin Seam Coal Mining Technology.* Noyes Technology Review No. 80. Park Ridge, NJ: Noyes Data, 1982. 385 p.

Literature survey, analysis and evaluation of mining systems. Problems are analyzed and compared; safety, production, and costs are considered. Design and costs for types of longwall mining systems are covered.

148. Colorado School of Mines. Mining Engineering Department. "Annotated Bibliography on Selected Mining Subjects." *Colorado School of Mines Quarterly,* Vol. 61, No. 2. p. 1–147.

Drilling and blasting, large-hole boring, grouting and reclamation are annotated briefly for the entries, which date from 1955–65.

149. Colorado State University. Natural Resources Ecology Laboratory. *The Ecological Effects of Coal Strip-Mining: A Bibliography with Abstracts.* Fort Collins, CO: Fish and Wildlife Service, 1977. 416 p. FWS/OBS-77/09.

Emphasis on western U.S. strip mining reclamation. References are grouped by topic, such as climate, plants and general reclamation. Very useful keyword-in-title index is included.

150. Economic Geology Publishing Co. *Annotated Bibliography of Economic Geology, 1929–1967.* Mount Vernon, MI, 1968.

Cumulative index useful for earlier works on mineral deposits.

151. Ellis, D., and Bercal, T. *Western Coal Development and Utilization; A Policy Oriented, Selected Bibliography with Abstracts.* McLean, VA: The Mitre Corporation, 1975. Various paging.

For the western coal states, policy-oriented bibliography with author-generated abstracts. Economics of coal development, water, pollution, legal, leasing and Indian aspects are emphasized. Entries are cross-referenced by location.

152. Erdmann, Charlotte A. *A Selective Guide to Literature on Mining Engineering.* Engineering Literature Guides, no. 6. Washington, DC: American Society for Engineering Education, 1985. 32 p.

Bibliography on mining engineering contains brief annotations for most entries. Sources of mining information are arranged by type of publication; indexes and databases are included.

153. Fejes, A. J., et al. *Subsidence Information for Underground Mines—Literature Assessment and Annotated Bibliography.* U.S. Bureau of Mines Information Circular IC 8990. Washington, DC: Government Printing Office, 1985. 86 p.

Pre-1984 literature is judged, assessed and summarized, and inclusion rated on applicability to certain areas. Subsidence problems and corrective measures are covered for an audience of mining industry personnel and regulatory authorities. A larger nonannotated list of entries is also included.

154. Fowkes, Richard S., and Mytrysak, Cynthia A. *Abstracts of Publications and Reports from Coal Mining Technology Research.* U.S. Bureau of Mines Open-File Report 122–85. Pittsburgh, PA: U.S. Bureau of Mines, 1985. 202 p. PB86-139003.
Abstracts of contracted research results in general, underground mining, surface mining, and mine systems, presented chronologically. Mining subtopics include planning, development, production systems, logistics and reclamation.

155. Hanslovan, James J., and Visovsky, Richard G. *Logistics of Underground Mining.* Energy Technology Review. Park Ridge, NJ: Noyes Data, 1984. 274 p.
Extensive cost data on transportation of personnel, equipment, coal and rock. Water handling is covered, as are communication systems.

156. Honkala, Rudolph A. *Surface Mining and Mined Land Reclamation; A Selected Bibliography.* Washington, DC: Old West Regional Commission, 1974. 154 p.
Comprehensive bibliography focuses on surface mining and reclamation in the western U.S., especially in Montana and Wyoming.

157. Kaplan, Stuart R. *Guide to Information Sources in Mining, Minerals, and Geosciences.* New York: Wiley, 1965. 598 p.
International coverage of literature and organizations concerning mining and minerals, arranged by area and country. Useful for earlier and continuing sources.

158. Lakos, Amos A., and Cooper, Andrew F. *Strategic Minerals: A Bibliography.* University of Waterloo Library Bibliography No. 14. Waterloo, ON: University of Waterloo Library, 1987. 132 p.
Resource politics of strategic minerals, those minerals necessary for industrial production, focus on post-1970 publications in this nonannotated English-language bibliography. Politics, economics, trade and cartel sections are followed by world geographic listings. Author and title index.

159. Mineral Information Institute. *Mineral Information Sources.* Denver, CO, 1982. 346 p.
Valuable source for films and booklets on the mineral industry. Classified by type of energy, environment, countries, and states with mineral topics. Sources of films are given, brief reviews, grade level of audience, and cost. Although the entries are ranked according to school grade level, many of the films would be useful for adult audiences also. An unusual and useful source.

160. Parkinson, George. *Guide to Coal Mining Collections in the United States.* Morgantown, WV: West Virginia University Library, 1978. 182 p.
Major coal mining collections are listed alphabetically, with address, hours, area of specialization, fees and photcopying availability. Archival collections are then listed and described, and a very detailed index by subject and person is included.

161. Peters, Robert H. *Foreign Literature on Environmental and Personal Factors Affecting the Safety and Productivity of Miners: A Topical Listing and Annotated Bibliography of Recent Research.* U.S. Bureau of Mines Open-File Report 193–83. Pittsburgh, PA: U.S. Bureau of Mines, 1983. 91 p. PB 84–127687.

Foreign literature in accident and safety, individual miner, health and medicine, technical and engineering, and ergonomics. Brief annotations for the 500 references which suggest workable solutions to common safety problems.

162. Ridge, John Drew. *Annotated Bibliographies of Mineral Deposits in Africa, Asia (Exclusive of the U.S.S.R.) and Australasia.* Oxford, New York: Pergamon, 1976. 545 p.

Survey articles on mineral deposits of specific provinces, plus extensive bibliographies. Good place to start research on a particluar area. Deposit location and characteristics are covered.

163. Ridge, John Drew. *Annotated Bibliographies of Mineral Deposits in Europe.* Part 1: Northern Europe Including Examples from the USSR in Both Europe and Asia. New York: Pergamon, 1984. 778 p.

Ore districts are mostly from Northern Europe, with only a few from USSR. Location, grade, tonnage, characteristics and some production are listed. Good summaries of mine ore characteristics and very good bibliographies are included for each one. All are written in English.

164. Ridge, John Drew. *Annotated Bibliographies of Mineral Deposits in the Western Hemisphere.* Geological Society of America Memoir 131. Boulder, CO: Geological Society of America, 1972. 681 p.

Overview of mineral deposits, their location, age, and Lindgren classification. Arranged by continent, then country and state or province for the U.S. and Canada. Summary of the deposit plus a list of references for further information. Author, deposit, and mineral indexes.

165. Simpson, Thomas A., and Phang, Michael K. *A Bibliography on Surface Mine Blasting.* U.S. Bureau of Mines Open-File Report 127-84. Pittsburgh, PA: U.S. Bureau of Mines, 1984. 41 p.

Noise and vibration problems from increased blasting in mines are the focus of this brief bibliography on mine blasting. Not available for purchase but can be consulted in selected U.S. Bureau of Mines libraries.

166. Veith, David L., et al. *Literature on the Revegetation of Coal Mine Lands: An Annotated Bibliography.* U.S. Bureau of Mines Information Circular 9048. Washington, DC: Government Printing Office, 1985. 296 p.

Extensive bibliography of U.S. and Canadian literature on surface coal mining and surface areas of underground mines, from 1977–84. Brief annotations with keyword access. Relevant agencies and other bibliographies are listed to increase access to revegetation literature.

167. Viola, John; Mack, Newell B.; and Stauffer, Thomas R. *Energy Research Guide; Journals, Indexes, and Abstracts.* Cambridge, MA: Ballinger, 1983. 284 p.

Over 500 periodicals in English, indexes and abstracts are covered with annotations. Alphabetical master list of titles, subject lists, and descriptions of periodicals.

168. Ward, Dederick C.; Wheeler, Marjorie W.; and Bier, Robert A., Jr. *Geologic Reference Sources; A Subject and Regional Bibliography of Publications and Maps in the Geological Sciences.* 2d ed. Metuchen, NJ: Scarecrow Press, 1981. 560 p.

Worldwide, but emphasis on North America, this relevant bibliography covers economic geology and environmental geology among its subjects. Useful especially for mineral deposits and maps included in the regional coverage. Some brief annotations.

169. Weber, R. David. *Energy Information Guide.* Santa Barbara, CA: ABC-Clio, 1982–84. 1,357 p. 3 vols.

Excellent inclusive bibliography on energy information, with substantive abstracts arranged according to type of energy: Volume I, General Alternative Energy Sources; Volume 2, Nuclear and Electric Power; Volume 3, Fossil Fuels. Author, title, subject and document number indexes aid in access to the 3,082 annotated entries.

170. Wilcox, Virginia Lee, comp. *Guide to Literature on Mining & Mineral Resource Engineering.* Washington, DC: American Society for Engineering Education, 1972.

Dictionaries, abstracts and indexes, and research centers are well covered in this literature guide.

171. Yanarella, Ernest J., and Yanarella, Ann-Marie. *Energy and the Social Sciences: A Bibliographic Guide to the Literature.* Westview Special Studies in Natural Resources and Energy Management. Boulder, CO: Westview Press, 1982. 347 p.

Guide to energy research and policies in the U.S. and worldwide. General perspectives, then specifics such as economics and fossil fuels in a bibliography without annotations. Also annotated bibliography of seventy-five essential energy books.

Indexes and Abstracts

Indexes and abstracts vary from those that cover primarily mining and mineral industries to those that index a small but significant area of the field. *Engineering Index* and *IMM Abstracts* offer relatively complete coverage, and the others index according to coverage noted. Virtually all the indexes are still published, and the three exceptions deal directly with mining and mineral economics.

172. *Applied Science & Technology Index.* Bronx, NY: H. W. Wilson Co., 1958– . (Monthly, with quarterly and annual cumulations)
This easy-to-use reference source indexes the primary mining journals relatively quickly, with specific index terms for easy retrieval. Previous title was *Industrial Arts Index.*

173. *Bibliography and Index of Geology.* Alexandria, VA: American Geological Institute, 1969– . (Monthly, with annual cumulations)
This standard index to geological literature covers economic geology and mineral deposits, engineering geology and its applications to mining, and newer issues provide more access to mines and mining districts. This index is by title only and does not include abstracts. The previous titles, *Bibliography and Index of North American Geology* and *Bibliography and Index of Geology Exclusive of North America,* provide indexing back to 1785 for North American geology and 1933 for non-North American information.

174. *Cadmium Abstracts.* London: Cadmium Association, 1977– . (Quarterly, with annual index)
Analyses, uses and environmental aspects of cadmium are indexed in this very specific set of abstracts.

175. *Chemical Abstracts.* Columbus, OH: Chemical Abstracts Service, 1907– . (Weekly, with semiannual index)
Mining is peripheral to this massive index, but explosives, geochemistry and pollution aspects of mining are indexed from journals, patents, conferences and monographs.

176. *Coal Abstracts.* London: International Energy Agency Coal Research, 1977– . (Monthly)
Information on all facets of coal production is abstracted, and abstracts which would have been published in the now defunct *Technical Coal Press Abstracts: Mining Technology* and its predecessors, *National Coal Board Abstracts A, B, C* and *D* are now incorporated in *Coal Abstracts.*

177. *Current Contents/Engineering, Technology & Applied Sciences.* Philadelphia, PA: Institute for Scientific Information, 1970– . (Weekly)
Tables of Contents of over 700 engineering journals and several hundred books are circulated, with keyword and author indexes.

178. *Dissertation Abstracts International. Section B: Physical Sciences and Engineering.* Ann Arbor, MI: University Microfilms, 1938– . (Monthly)
Dissertations are abstracted with fairly lengthy descriptions, arranged under general subjects, with UMI order numbers attached to each entry. Semiannual indexes provide more specific subject access, plus author access.

179. *EI Engineering Conference Index.* New York: Engineering Information, 1985– . (Monthly)
Papers presented at engineering conferences worldwide are indexed using the same thesaurus terms as *Engineering Index.*

180. *Energy Abstracts.* New York: EIC/Intelligence, 1970– . (Monthly, with annual cumulation)
Coal and uranium mining information is abstracted from journal articles, technical reports and some government documents, mostly concerning the U.S. and Canada. Abstracts are brief but informative, and microfiche of the entire text of each reference are available for most items.

181. *Energy Bibliography & Index.* Houston, TX: Gulf Publishing, 1978– . (Irregular)
This comprehensive guide to energy materials in the Texas A&M University Library contains brief abstracts and indexes by subject, keyword-in-title, author, corporate author and report series. All types of published information are covered except for journal articles.

182. *The Energy Index.* New York: EIC/Intelligence, 1970– . (Annual)
Index to *Energy Abstracts* contains fairly general index terms which locate mineral industry and mining information.

183. *Energy Research Abstracts.* Washington, DC: Government Printing Office, 1976– . (Monthly, with annual cumulations)
This comprehensive index covers all scientific and technical reports and articles, patents, theses and monographs by the Department of Energy staff and contractors. Abstracts are brief and descriptive, and index terms include personal author, corporate author, contract number and report number as well as subject.

184. *Engineering Index.* New York: Engineering Information, 1884– . (Monthly, with annual cumulations)
This basic engineering index has great breadth and depth of coverage on mining engineering and mineral industries from worldwide sources, including extensive coverage of Russian literature. Abstracts are brief but substantive, and all are in English.

185. *Environment Abstracts.* New York: EIC/Intelligence, 1971– . (10/yr, with annual cumulation)
Environmental aspects of mining, mine waste management, and pollution are indexed from journals, conferences and technical reports. The brief abstracts emphasize the U.S. and Canada but do include other areas. Microfiche of the complete text are available for most items.

186. *Environment Index.* New York: EIC/Intelligence, 1971– . (Annual)
General index terms are used to access the items abstracted in *Environment Abstracts.* The index can be used alone, without the abstracts, as access to journal articles and other materials on the environment.

187. *Fuel and Energy Abstracts.* Guildford, England: Butterworth, 1960– . (Bimonthly)
Published on behalf of the Institute of Energy, these abstracts provide comprehensive coverage of world literature on all aspects of fuel and energy, including coal and nuclear, with very specific subject index and author access in each issue. Some abstracts are secondary from other sources.

188. *Geomechanics Abstracts.* In *International Journal of Rock Mechanics and Mining Sciences & Geomechanics Abstracts.* Elmsford, NY: Pergamon, 1964– . (Bimonthly)
Concise abstracts of technical information on rock mechanics, including aspects applicable to mining, are arranged by general subject and cover the world's literature on rock mechanics. All abstracts are in English.

189. *Government Reports Announcements & Index.* Springfield, VA: National Technical Information Service, 1965– . (Monthly, with annual index)
This overall index to government-sponsored research covers mining and processing reports submitted by government contractors. Index points are keyword, personal author, corporate author, contract number, and NTIS order number. U.S. Bureau of Mines Open-File Reports are often indexed here.

190. *HSL Abstracts.* Sheffield, England: Health and Safety Laboratories, 1951– . (Bimonthly, with annual index)
Materials in many languages from many countries are abstracted in English and arranged according to Universal Decimal Classification, with annual subject and author indexes. Previous title was *Safety in Mines Abstracts.*

191. *IMM Abstracts.* London: The Institution of Mining and Metallurgy, 1950– . (Bimonthly)
Focusing on mining, mineral exploration and processing of minerals except coal, the IMM Library staff abstract journals, technical reports, conference proceedings and books received, and prepare an annual index to the abstracts. Copies of the items are available from the IMM Library, and order forms are included.

192. *Index of Mining Engineering Literature.* New York: Wiley, 1909–1912. 2 vols.
This very early index to literature in mining engineering provides historical access to early mining practices and emphasis.

193. *Index to Scientific and Technical Publications in Print.* Philadelphia, PA: Institute for Scientific Information, 1986– . (Monthly, with annual cumulations)
Covers papers, not abstracts, from published proceedings, including preprints and journals. Geosciences and engineering are included, and mining is a subject entry, but a minor part of the index. Nevertheless, valuable index for access to conference papers not indexed elsewhere. Some overlap with *EI Engineering Meetings*.

194. *INIS Atomindex.* Vienna, Austria: International Atomic Energy Agency, 1970– . (Monthly, with semiannual cumulation)
International coverage of uranium, thorium and vanadium mining, ores and compounds. Uses of nuclear materials and status of the nuclear industry are also reviewed.

195. *KWIC Index to Rock Mechanics Literature, Part 2, 1969–1976.* New York: American Institute of Mining, Metallurgical and Petroleum Engineers, 1979. 731 p.
Keyword index to the world's rock mechanics literature during this period provides easy access to bibliographic listings, and an author index is also included.

196. *KWIC Index to Rock Mechanics Published before 1969.* New York: American Institute of Mining, Metallurgical and Petroleum Engineers, 1971. 2 vols.
Keyword indexing of early rock mechanics literature from 1870–1968 provides easy retrieval of materials on very narrow topics as well as on broader ones. An author index is also included.

197. *Lead Abstracts.* London: Lead Development Association, 1958– . (Quarterly)
Specialized coverage of lead and its alloys, from analysis to uses and environmental impacts is accomplished in short abstracts from worldwide sources.

198. *Masters Abstracts.* Ann Arbor, MI: University Microfilms, 1962– . (Quarterly)
Masters theses from participating organizations are abstracted, with subject and author indexes.

199. *Master's Theses in the Pure and Applied Sciences.* West Lafayette, IN: Purdue University, 1955– .
Arranged by departmental major and then by university, this index lists authors and titles of mining-engineering theses at participating institutions. No subject index, aside from the title of the department, is provided, so scanning list of theses is necessary.

200. *Metals Abstracts.* Metals Park, OH: American Society for Metals, 1968– . (Monthly, with annual cumulation)
Worldwide technical literature of physical properties of metals, their ores, processing and uses is covered by descriptive abstracts. Indexing is exhaustive, specific and relevant. Earlier titles were *Review of Metal Literature* and *Metallurgical Abstracts.*

201. *Mineral Economics Abstracts.* New York: ATLANTIS Energy and Mineral Economic Services, Inc., 1978–81. (Monthly)
Abstracting service translated from *Revue De Presse d'Economic Miniere.* Worldwide coverage with brief abstracts that summarize salient points of the publication.

202. *Mineralogical Abstracts.* London: Mineralogical Society, 1959– . (Quarterly)
The world literature of mineralogy is covered by abstracts in English, with indexing to specific terms.

203. *Mining Technology Abstracts.* Ottawa, ON: Technology Information Division, CANMET, Division of Energy, Mines and Resources Canada, 1983– . (Biweekly)
Good coverage of rock mechanics and mining technology information, especially concerning hard-rock mining, is contained in concise abstracts with indexes for specific access.

204. *Mining World Index of Current Literature.* Chicago: Mining World Company, 1912–1916. (Semiannual)
Early literature of mining and metallurgy was indexed by broad topic and subdivided into narrower categories. The index continued an earlier index published in the 1911 issues of *Mining and Engineering World.* The index is now useful primarily for historical information to trace trends or compare practices.

205. *Monthly Catalog of United States Government Publications.* Washington, DC: Government Printing Office, 1951– . (Monthly, with annual cumulation)
This index and listing of government documents from all government agencies includes those of the Environmental Protection Agency, Office of Technology Assessment and other relevant agencies. Because of its broad coverage, this index is useful in addition to the more specific indexes of the Bureau of Mines and U.S. Geological Survey.

206. *New Publications of the U.S. Geological Survey.* Washington, DC: Government Printing Office, 1962– . (Monthly)

   Subject and author access to new U.S. Geological Survey (USGS) publications is cumulated annually into *Publications of the U.S. Geological Survey.* This essential index is very useful for mineral deposits, mining districts, mines, particular commodities and ground-water information. For mining information, it is a logical companion to the U.S. Bureau of Mines index series.

207. *New Silver Technology.* Washington, DC: The Silver Institute, 1974– . (Quarterly)

   For those interested in the silver industry, *New Silver Technology* contains page-long abstracts on journal articles, international patents, industry and government reports on worldwide technology and uses of silver. The lengthy abstracts contain the most information on specific topics to be found in all the mining and mineral-industry abstracts.

208. *Oceanic Abstracts.* Bethesda, MD: Cambridge Scientific Abstracts, 1964– . (Bimonthly)

   Ocean mining and dredging, particularly for manganese nodules, are abstracted and indexed, but this is a minor part of the overall index.

209. *PAIS Bulletin.* New York: Public Affairs Information Service, 1915– . (Monthly, quarterly or annual, depending on type of membership)

   Journal articles, books, government documents and conference proceedings are indexed in English. Mineral industries and mining are indexed if policy-oriented, such as mineral rights, legislation, and the role of mining industries in governmental policy.

210. *Petroleum Abstracts.* Tulsa, OK: University of Tulsa, 1961– . (Weekly, with annual index)

   These abstracts provide substantial information in brief annotations of information sources on commodities such as gold, copper and silver, especially in the United States. Although the overall focus is on petroleum, considerable information on minerals is included. Very specific and direct access points, so the index is easy to use.

211. *Pollution Abstracts.* Bethesda, MD: Cambridge Scientific Abstracts, 1970– . (Bimonthly)

   Pollution from mining, metals, mine waste and mine drainage are abstracted, with index giving very specific access. Mining and minerals are a minor part of the overall coverage.

212. *Predicasts F&S Index Europe.* Cleveland, OH: Predicasts, 1979– . (Monthly, with quarterly and annual cumulations)

   Newspapers and business journals are indexed by type of mining and by mining company name for information on current activities, planned mines, expansions and other company data from European mining firms. Although mining and mineral industries are a minor part of the index, the information covered still offers valuable insights into company activities.

213. *Predicasts F&S Index International.* Cleveland, OH: Predicasts, 1967– . (Monthly, with quarterly and annual cumulations)
Worldwide mining activities exclusive of the United States and Europe are included in this index by type of industry, country, and company name. As in the other Predicasts indexes, company activities and those of a particular type of commodity, such as gold mining, can be ascertained using this index.

214. *Predicasts F&S Index United States.* Cleveland, OH: Predicasts, 1960– . (Monthly, with quarterly and annual cumulations)
This index is very useful to track current activities of mining companies, either by type of mining or by company name. It is especially valuable for pinpointing small mining companies or subsidiaries of larger corporations where information is sparse and diverse.

215. *Publications of the U.S. Geological Survey.* Washington, DC: Government Printing Office, 1879– . (Annual, with periodic cumulations)
Subject and author access to USGS publications is very useful for mineral deposits, mining districts, and particular commodities. Cumulations include 1879–1961, 1962–70, and 1971–81.

216. *Science Citation Index.* Philadelphia, PA: Institute for Scientific Information, 1967– . (Quarterly, with annual cumulations)
Citations to previously written materials can be checked through this index, which includes a wide range of scientific journals and books. Related work done recently can be found through the citations to earlier materials. Subject access is available through permuterm index, and author through source and citation volumes. This index is one of the fastest to include current materials.

217. *World Aluminum Abstracts.* Washington, DC: Aluminum Association, 1968– . (Monthly)
The brief abstracts cover physical properties and uses of aluminum and its alloys.

218. *Zentralblatt Fuer Mineralogie, Teil I: Kristallographie, Mineralogie.* Stuttgart, West Germany: E. Schweizerbart'sche Verlagsbuchhendlung, 1807– . (7/yr)
Many English titles and abstracts on mineralogy and crystallography are included, as well as the German. Titles in languages other than English or German are given in German, with the original language noted.

219. *Zentralblatt Fuer Mineralogie, Teil II: Petrographie, Technische Mineralogie, Geochemie und Lagerstaettenkunde.* Stuttgart, West Germany: E. Schweizerbart'sche Verlagsbuchhenglung, 1807– . (13/yr)
Many English titles and abstracts on petrology, geochemistry and related subjects are included. Titles other than English or German are listed in German and the original language noted.

220. *Zinc Abstracts.* London: Zinc Development Association, 1943– . (Quarterly)

Brief abstracts of worldwide publications, especially conference proceedings and journals, are indexed for technology and uses of zinc and its alloys.

Databases

Computerized databases supply easy, quick access to bibliographic and textual information through interactive online searching on specific topics. Mining and mineral industries information is available on many databases, either as the main focus (a few), or as a portion of a much larger set of information (many). Because searches can be tailored to a very specific retrieval, the many databases are listed and their mineral coverage noted. Print equivalents of the databases are noted. Some databases require subscription, such as the online newsletter files, and some vendors charge only for time online and references retrieved. Either way, online searching is an integral part of research today.

221. Alberta Oil Sands Index. Edmonton, AB: Alberta Oil Sands Information Centre, Alberta Research Council, 1900– . Available through CAN/OLE, QL Systems.
    Permission from the producer is required to access this small, focused database on Alberta oil sands, their geology, economics, mining and production.

222. Applied Science & Technology Index. Bronx, NY: H. W. Wilson, 1983– . Available through WILSONLINE.
    Reference index covers mining and mineral industries with same specific subject access as print version. Relatively fast indexing of U.S., British, Canadian, Dutch, Swiss and Irish journals in science and technology. Drawback is short time coverage, only from November 1983. Updated twice a week.

223. Coal Data Base (Coal Abstracts). London: IEA Coal Research, 1978– . Available through CAN/OLE, QL Systems Limited.
    Worldwide coverage of coal industry through indexing and abstracting journal articles, reports, conference proceedings. Coal exploration, mining, preparation, properties, and waste management are some of the topics. Computerized version of *Coal Abstracts*. Updated monthly.

224. Compendex. New York: Engineering Information, Inc., 1969– . Available through BRS, CAN/OLE, DIALOG, ORBIT, Pergamon, STN International.
    The online version of *Engineering Index Monthly* includes citations and abstracts of mineral industry costs, management, minerals availability,

mining and technology, excluding patents. Probably the best database for mining and mineral industry information, with very good coverage of conferences, the individual papers indexed in *EI Engineering Meetings.* Updated monthly.

225. DOE Energy. Washington, DC: U.S. Department of Energy, Office of Scientific and Technical Information, 1974– . Available through DIALOG, Mead Data Central, STN International.

Citations and abstracts to unclassified technical information from the DOE Technical Information Center in Oak Ridge. Covers geosciences, coal, nuclear energy, oil shale and other types of energy. Covers U.S. Bureau of Mines energy publications since 1910, and other entries earlier than 1974.

226. EBIB. College Station, TX: Texas A&M University Library, 1919– . Available through Orbit.

Citations on energy publications in the Texas A&M Library are indexed and abstracts provided. This online version of *Energy Bibliography and Index*, like its print counterpart, covers all types of literature except journals. All types of fuels are indexed.

227. Ecomine. Paris: Bureau de Recherches Geologiques et Minieres, 1984– . Available through Telesystemes-Questel.

Bibliographic citations and abstracts on the worldwide mineral industry in English and French. Financial information, reserves and production and activities of mining companies. Updated monthly.

228. EI Engineering Meetings. New York: Engineering Information, Inc., 1982– . Available through CAN/OLE, DIALOG, ORBIT, Pergamon InfoLine, STN International.

The online version of *EI Engineering Meetings* is an essential counterpart of Compendex, which contained the conference papers information before 1982. Engineering conferences and symposia around the world are covered with a main record and papers indexed individually. Some conferences from 1979–81 are also included. Abstracts were added in 1985. Updated monthly.

229. Energyline. New York: EIC/Intelligence, Inc., 1971– . Available through DIALOG, ESA-IRS; ORBIT.

Energy-related reports, conferences and other literature are indexed and abstracted in the online version of *Energy Index* and *Energy Information Abstracts.* Useful for economics, government and policy aspects of coal and nuclear energy. Updated monthly.

230. Enviroline. New York: EIC/Intelligence, Inc., 1971– . Available through DIALOG, ESA-IRS; ORBIT.

Environmental impacts of mining and uses of mineral resources are a small but useful portion of this online version of *Environment Index* and *Environment Abstracts.* Coverage is international and reflects concerns and policies about minerals. Updated monthly.

231. Environment Reporter. Washington, DC: The Bureau of National Affairs, 1982– . Available through Mead Data Central.
Full text of Current Developments section of *Environment Reporter,* covers legal, political, enforcement and general news about the environment and current impacts. Updated monthly.

232. Ferroalloys and Strategic Metals Forecast. Bala Cynwyd, PA: Chase Econometrics, 1970– . Available through Chase Econometrics.
Numeric database with ten-year projected supply, demand, and costs for metals. Updated twice a year.

233. Geoarchive. Geosystems, 1974– . Didcot, Oxon, England: Available through DIALOG.
Bibliographic database with good international coverage of mineral resources and geology sources in the British Geological Survey. Mining coverage better in the 1970s than later. Updated monthly.

234. Geode. Paris: Bureau de Recherches Geologiques et Minieres, 1977–1983. Available through Telesystems—Questel.
Bibliographic database covering world literature on the mining industries from 1977–83, in English and French. Later information in *Pascal: Geode.*

235. Geomechanics Abstracts. London: Rock Mechanics Information Service, Imperial College of Science and Technology, 1977– . Available through Pergamon InfoLine.
Rock mechanics abstracts online correspond to print *Geomechanics Abstracts* in *International Journal of Rock Mechanics and Mining Sciences and Geomechanics Abstracts.* The database updates the *KWIC Index to Rock Mechanics* and continues the excellent keyword access. Technical database updated every two months.

236. GeoRef. Alexandria, VA: American Geological Institute, 1785– . Available through CAN/OLE; DIALOG, ORBIT.
Bibliographic database of worldwide geologic and minerals information. The length of coverage makes this a unique and valuable online resource; coverage extends to 1785 for North America and is being extended for the rest of the world. Lag time between publication and indexing ranges from several months to several years. Information indexed includes journals, conferences, government publications, theses and dissertations. Online version of *Bibliography and Index of Geology* and earlier variations. Updated monthly.

237. GPO Monthly Catalog. Washington, DC: U.S. Government Printing Office, 1976– . Available through BRS, DIALOG, TECH DATA.
Government publications are indexed in this online equivalent of *Monthly Catalog of United States Government Publications.* U.S. Bureau of Mines and U.S. Geological Survey publications are covered, but the relatively short period of coverage limits its effectiveness for mining and mineral industry information. Updated monthly.

238. HSELINE. Sheffield, England: Health and Safety Executive, Library and Information Services, 1977– . Available through DATA-STAR, ESA-IRS, Pergamon InfoLine.
Bibliographic database on worldwide safety and health in mining, explosives, nuclear technology and industrial pollution covers British books, journals, technical reports, and legislation. Materials from other countries are also included, with English abstracts for non-English materials. Updated monthly.

239. IMMAGE: Information on Mining, Metallurgy and Geological Exploration. London: Library and Information Services, The Institution of Mining and Metallurgy, 1979– . Available through IMM.
Comprehensive bibliographic database on worldwide economic geology, mining, nonferrous metals and industrial minerals. Updated monthly.

240. International Energy Annual. Rochester, NY: I. P. Sharp, 1973– . Available through I. P. Sharp.
Text and numeric database of annual time series on supply, prices, imports and exports for coal and nuclear energy. Online version of *International Energy Annual.* Updated annually.

241. International Lead and Zinc. Bala Cynwyd, PA: Chase Econometrics, 1955– . Available through Chase Econometrics by subscription.
Text and numeric time series on lead and zinc production, uses and trade worldwide. Updated monthly.

242. Iron and Steel. Bala Cynwyd, PA: Chase Econometrics, 1972– . Available through Chase Econometrics by subscription.
Iron and steel production and consumption in the U.S. and internationally. Updating varies.

243. Magnesium Forecast. Bala Cynwyd, PA: Chase Econometrics, 1970– . Available through Chase Econometrics by subscription.
Twelve-year forecasts for supply, demand, production and prices for magnesium. Updated twice a year.

244. Marine Minerals Data Base. Boulder, CO: National Geophysical Data Center, National Oceanic and Atmospheric Administration, 1983– . Available only as customized searches, not directly online.
Over 10,000 items, U.S. and foreign, including books, journal articles, workshop proceedings, lab reports and theses on marine minerals. Dual database of data and bibliographic references, with geochemical data, searchable by latitude/longitude and water depth.

245. MILS: The Mineral Industry Location System. Denver, CO: U.S. Bureau of Mines Intermountain Field Operations Center. Available through U.S.B.M.
Searches can be requested on this database, which contains information on mineral deposits, developed or prospective. Basic data and reference information are included. Not currently available in interactive mode.

246. Minerals Data System. University of Oklahoma, 1972–. Available through General Electric Information Services Co.

Originally developed by the U.S. Geological Survey, the Computerized Resources Information Bank (CRIB) contains basic data on mineral resources from USGS investigations. A Mining District File has also been added.

247. Minesearch. Metals Economics Group, Ltd, 1982–. Available on Martin Marietta Data Systems by subscription to Metals Economics Group.

Full-text database on more than 75,000 U.S. mineral deposits, with sample and core data, production, reserves, costs, and sources of information. Very good source of information. Updated irregularly.

248. Minproc. Energy, Mines & Resources Canada, Centre for Mineral and Energy Technology (CANMET), 1978– . Available through QL Systems Limited.

Bibliographic database on mineral processing, primarily from Canadian publications in English, but also some international coverage. Unusual searching algorithm. Updated biweekly.

249. MinSys. Didcot, Oxon, England: Geosystems, 1980– . Available through Geosystems.

Very large database, with over five million records on worldwide mining and minerals industries. Exploration, production, transportation, management, and marketing are all covered in this bibliographic and text file. Updated daily.

250. Mintec. Ottawa, ON: Energy, Mines & Resources Canada, Centre for Mines and Energy Technology (CANMET), 1968– . Available through QL Systems Limited.

Relatively small, focused bibliographic database especially useful for hard-rock mining technology. Especially strong on Canadian literature, including government studies, but also includes some international coverage. Some entries are in French. Unusual searching algorithm. Updated biweekly.

251. Nexis Magazines. Dayton, OH: Mead Data Central, Inc., 1981– . Available through Mead Data Central, Inc. by subscription.

Full text online of four premium mining publications, *Mining Journal, Mining Magazine, E&MJ Engineering and Mining Journal* and *Mining Annual Review,* along with many other magazines. Can be searched by journal or by a subject group. Updated according to frequency of journal publication.

252. Nexis Newsletters. Dayton, OH: Mead Data Central, Inc. Available through Mead Data Central by subscription.

Full text online of several important mining newsletters, *E&MJ Mining Activity Directory,* 1982–; *Keystone News Bulletin,* 1982–; and *Metals Week,* 1981–, as well as many other newsletters. Can be searched by journal name or subject group. Updated according to frequency of publication.

253. NTIS. Springfield, VA: National Technical Information Service, 1964– . Available through BRS, CAN/OLE, DATA-STAR, DIALOG, ESA-IRS, Knowledge Index, Mead Data Central, ORBIT, STN.
Online version of *Government Reports Announcements*. Bibliographic database, including abstracts, of government-funded research on science and technology, including mining and mineral industries. Single-word indexing sometimes produces unusual retrieval of references, but still a good source for mining, as U.S. Bureau of Mines Open-file reports are covered. Updated monthly on most vendors, bimonthly on others.

254. Oceanic Abstracts. Bethesda, MD: Cambridge Scientific Abstracts, 1964– . Available through DIALOG, ESA-IRS.
Marine mining is one of the subjects covered in this computerized version of the print *Oceanic Abstracts*. Updated bimonthly.

255. PAIS International. New York: Public Affairs Information Service, Inc., 1976– . Available through BRS, DATA-STAR, DIALOG, Tech Data.
Public policy database with references to economics and other social sciences. Mineral industries are covered in both the English language and foreign entries, which correspond to *PAIS Bulletin* and *PAIS Foreign Language Index*. Updated monthly for English-language, quarterly for foreign-language.

256. Pascal: Geode. Paris: Centre National de la Recherche Scientifique, Centre de Documentation Scientifique et Technique, 1977– . In French, with titles and descriptors also in English. Available through ESA-IRS; Telesystemes Questel.
Earth science database with information on mineral economics, mineral deposits and mineralogy, as well as the broader earth sciences. Often citations are not available in other databases, so citations in English can be retrieved. Updated monthly.

257. PTS Annual Reports Abstracts. Cleveland, OH: Predicasts, Inc. Available through BRS, Data-Star, DIALOG, TECH DATA.
Current reports with up to five years of back data. Mineral industry companies report activities, reserves, and plans, which are then available in this textual-numeric database. Useful to track performance of particular company.

258. PTS F&S Indexes. Cleveland, OH: Predicasts, Inc., 1972– . Available through BRS, Data-Star, DIALOG, TECH DATA.
Excellent bibliographic access to mineral industry and mining activities around the world, with access by company name and type of industry. Especially useful for companies which are subsidiaries but will be listed under their own name in the index. Corresponds to *Predicasts F&S Index United States, Predicasts F&S Index Europe,* and *Predicasts F&S Index International*. Updated weekly on DIALOG, monthly on the other vendors.

259. PTS International Forecasts. Cleveland, OH: Predicasts, Inc., 1971– . Available through DIALOG.
Usually contain base data, short-term forecast and long-term forecast for mineral commodity such as platinum. Source of data is included in this useful database. Updated monthly.

260. PTS U.S. Forecasts. Cleveland, OH: Predicasts, Inc., 1971– . Available through DIALOG and through DATA STAR as one database combined with *PTS International Forecasts.*
As with the international forecasts, base data and forecasts for mineral commodities are covered, and sources are included. Updated monthly.

261. PTS U.S. Time Series. Cleveland, OH: Predicasts, Inc., 1973– . Available through DATA STAR, DIALOG.
Annual time series of U.S. economic data on production and use of mineral commodities, among others. Very useful for quick information, and sources lead to further data. Some series include data from 1957. Updated quarterly.

262. Science Citation Index. Philadelphia, PA: Institute for Scientific Information, 1967– . Available through DIALOG, ORBIT.
Based on citation analysis through scientific journals and books. Can be searched by item cited, author, source of citation, or permuterm subject. Very quick indexing. Updated monthly.

263. Thomas Register Online. New York: Thomas Publishing Company, Inc. Available through DIALOG.
Reference database on U.S. manufacturers and service companies includes name, address, and products. Although mining is a small part of the database, equipment manufacturers can be located through this online version of part of *Thomas Register of American Manufacturers.* Updated annually.

264. Tulsa. Tulsa, OK: Petroleum Abstracts, 1965– . Available through ORBIT.
Online version of *Petroleum Abstracts,* it contains bibliographic references and abstracts of mineral commodity information as a minor but significant portion of the database. Updated weekly.

265. U.S. Cost Planning Forecast. Bala Cynwyd, PA: Chase Econometrics, 1947– . Available through Chase Econometrics by subscription.
Quarterly and annual time series of historical and forecast data for metals and metal products as part of the data provided. Updated monthly for short-term forecasts, quarterly for long-term.

266. U.S. Energy Forecast. Bala Cynwyd, PA: Chase Econometrics, 1960– . Available through Chase Econometrics by subscription.
Quarterly and annual historical and forecast data on energy consumption and prices of coal and nuclear power, along with the petroleum energy sources. Useful for comparisons among competing sources of power. Updated quarterly.

267. World Aluminum Abstracts. Metals Park, OH: American Society for Metals, 1968– . Available through DIALOG, ESA-IRS.
Bibliographic index to literature for the entire range of the aluminum industry from all types of sources. Updated monthly.

268. Zinc, Lead & Cadmium Abstracts. London: Zinc Development Association, 1970– . Available through Pergamon InfoLine.
References, with abstracts, to production, characteristics and uses of zinc, lead, cadmium, and their alloys from patents, journals, and conference proceedings. Online version of the print abstracts on each commodity with additional entries. Updated monthly.

Industry Surveys

These surveys of the mineral industry as a whole, or a portion of the industry, offer statistical or analytical reviews. Mining in the developing countries has drawn considerable attention in the 1970s and 1980s, as has gold, and several surveys reflect those interests.

269. American Iron and Steel Institute. *Annual Statistical Report.* Washington, DC, 1879–.
In addition to iron and steel production, imports and basic materials for steel are summarized with five years of statistics. Industry trends can be followed from the data, and current production evaluated.

270. American Iron Ore Association. *Iron Ore.* Cleveland, OH, 1957– . (Annual)
Iron-ore industry of the world, but focused on the U.S. and Canada, is reported in tabular form of iron-ore production and shipments. Imports and inventory by area are tabulated, and a directory of reporting companies completes the data.

271. Bosson, Rex, and Baron, Bension. *The Mining Industry and the Developing Counties.* New York: Oxford University Press for the International Bank for Reconstruction and Development/The World Bank, 1977. 292 p.
Valuable study of the problems of mineral mining in the developing countries, from exploration through development and trade. Suggests a mining code. Canadian data, analyzed because of their availability, are extrapolated for developing countries.

272. Boyle, E. H., Jr.; Peterson, G. R.; and Thomas, P. R. *Primary Silver Availability—Market Economy Countries.* A Minerals Availability Appraisal. U.S. Bureau of Mines Information Circular 9090. Washington, DC: Government Printing Office, 1986. 36 p.
Silver from 436 mines and deposits in forty-one market economy countries is assessed for cost and availability. Silver ores and zinc, lead, copper and gold deposits with silver components were studied. At current prices, silver as a byproduct or coproduct is more favorable than that from predominantly silver ores. Interesting judgment on the commodity of silver.

273. Canada Department of Energy, Mines and Resources. *Canadian Minerals Yearbook.* Ottawa, ON: 1962– . (Annual)
Minerals activity in Canada is summarized each year, with indexing by company. Useful overview of Canadian mining and mineral production.

274. Coal Industry Advisory Board. *The Use of Coal in Industry.* Paris: International Energy Agency, 1982. 445 p.
Coal use in industry is projected to the year 2000 and compared to oil, considering a wide price difference between the two. Geared toward increasing coal use in industry, this study considers costs of conversion to coal. Although projections of costs have not held true, the conversion information is still valid.

275. Connell, Norman, ed. *World Minor Metals Survey.* 2d ed. Surrey, England: Metal Bulletin PLC, 1981. 118 p.
Excellent comprehensive survey of trading, markets, and prices in the commercially active minor metals. A directory of producers, by country; and a directory of products, by location, capacity, form and grade add to the usefulness. An index to traders completes the survey.

276. Coope, B. M., and Dickson, E. M., eds. *Raw Materials for the Refractories Industry.* An Industrial Minerals Consumer Survey. London: Metal Bulletin, 1981. 128 p.
Worldwide supply and demand for industrial minerals used in refractories. An overview of the industry, then survey by mineral, plus buyer's guide. A good survey of an often overlooked minerals market.

277. Cunningham, Simon. *The Copper Industry in Zambia; Foreign Mining Companies in a Developing Country.* New York: Praeger, 1981. 343 p.
Exploration and production policies, and financial factors are examined, not politics as such. Specific mines illustrate the financial realities of private development, and the partial nationalization in 1969 is analyzed for impact on the copper industry. Useful as a possible pattern for multinational mining in Third World countries.

278. DeWit, Maarten J. *Minerals and Mining in Antarctica; Science and Technology, Economics and Politics.* Oxford: Clarendon Press, 1985. 127 p.
Survey of mineral deposits and feasibility analysis of a typical mining project in Antarctica, plus legal ramifications in this scholarly technical work. Bibliography and index.

279. Economic Assessment Service. *The Future Economics of Coal-Based Energy in the Residential Market.* EAS Report No. H3182. London: International Energy Agency, 1983. 175 p.
Increased use of coal will require changes in production, transportation and use. Coal use for synthetic natural gas in residential heating is examined, and displacement of oil and gas around 2010 is predicted.

280. Frisch, J. R., ed. *Energy 2000-2020: World Prospects and Regional Stresses.* Translated by P. Ruttley. World Energy Conference. London: Graham & Trotman, 1983. 259 p.

Interesting forecasts of energy production and consumption, and the interrelationship of coal, petroleum and nuclear power. Coal and nuclear power will have increased demand, according to the ten regional teams who produced the forecasts. Two scenarios are considered: cooperation between nations and international tension between nations. Either way, demand will increase.

281. Grossling, Bernardo. *World Coal Resources.* London: Financial Times, 1979. 164 p.

The pattern of coal supply, overview of coal occurrences, technology, and impediments to production are examined. The production and consumption of coal internationally is compared to other fuels.

282. Gupta, Satyadev. *The World Zinc Industry.* Lexington, MA: Lexington Books, 1982. 203 p.

This survey describes market structure for zinc, develops an econometric model which closely simulates actual consumption, and interprets institutional and market forces which influence the worldwide zinc industry.

283. Inga, Max. *A Model for Economic Evaluation of Alternative Mineral Development Strategies in Peru.* Unpublished Ph.D. dissertation, Colorado School of Mines, 1982. 378 p.

Case study of nonfuel mineral development in Peru, through analysis of Peruvian economic and political characacteristics, especially mining, from 1950–81. Importance of factors on industry development were ranked by Peruvian mineral-industry personnel through interviews, so value of various strategies to society could be evaluated through computer analysis. Alternative development strategies are presented, evaluated, and preferred development strategy analyzed and justified.

284. International Tin Council. *World Tin Mining; Operations, Exploration and Developments.* London, 1982. 86 p.

World tin mining is summarized in tables listing country, company, location, operations, output, future plans, and source of data. Useful for overall picture of the tin mining industry.

285. James, Peter. *The Future of Coal.* London: Macmillan Press, 1982. 271 p.

Based on assumptions of economic growth, world coal trade by area is analyzed, and coal industries and technologies worldwide are assessed. Technology and markets are the focus.

286. Kimbell, Charles L.; Lyday, Travis Q.; and Newman, Harold R. *Mineral Industries of Australia, Canada and Oceania, Including a Discussion of Antarctica's Mineral Resources.* U.S. Bureau of Mines Mineral Perspective. Washington, DC: Government Printing Office, 1985. 70 p.

Commodity production, reserves, trade and major mining companies are reviewed. Accompanying maps show locations of deposits, mines and processing plants. Very useful recent overview for the area.

287. LeFond, Stanley J. *Handbook of World Salt Deposits.* New York: Plenum, 1969. 384 p.
Study of salt deposits around the world. An excellent survey of deposits, their locations and characteristics.

288. Mangone, Gerard J., ed. *American Strategic Minerals.* New York: Crane Russak, 1984. 153 p.
Strategic minerals, those critical to American security, are examined for U.S. dependence, future demands, investments in developing countries, and politics. The need for a coherent minerals policy is stressed. Index included.

289. Marsh, S. P.; Kropschot, S. J.; and Dickinson, R. G., eds. *Wilderness Mineral Potential: Assessment of Mineral Resource Potential in U.S. Forest Service Lands Studied 1964–1984.* U.S. Geological Survey Professional Paper 1300. Washington, DC: Government Printing Office, 1984. 2 vols., 1,183 p.
U.S. Geological Survey and U.S. Bureau of Mines geologists and mining engineers evaluated present and potential wilderness areas for mineral resources. Character and geologic setting of each area are briefly described, resource potential assessed, and a reference list provided. Indexed by wilderness area.

290. Martino, Orlando; Hyde, Doris; and Velasco, Pablo. *Mineral Industries of Latin America.* U.S. Bureau of Mines Mineral Perspectives. Washington, DC: Government Printing Office, 1981. 117 p.
Summary of thirty-four countries and areas. Reserves, production, trade and principal mining countries are discussed. Useful as locator for deposits and as directory of major mining companies in Latin America.

291. Metal Bulletin Ltd. *World Aluminum Survey.* 5th ed. Edited by Norman Connell. Surrey, England, 1981. 278 p.
This excellent survey covers demand, trends, and recycling. Producers are listed by country, with address, ownership and products. Products and operations are reviewed by product, such as bauxite, and primary and secondary operations are included.

292. Metal Bulletin Ltd. *World Copper Survey 1980.* Surrey, England: Metal Bulletin, 1980. 210 p.
This survey resulted from questionnaires sent to companies active in copper mining and production. A directory by country lists head office, revenue, and plants in the copper industry. In addition, articles on copper agreements, market regulation, trends, recycling and type of mining are included. Another very good worldwide survey.

293. Muir, W. L. G. *Review of the World Coal Industry to 1990.* Wembley, England: Miller Freeman, 1975. 134 p. 1976 Supplement, 36 p.

A survey by country was published just after oil prices jumped and coal became a logical alternative fuel. The focus is electricity generation. Rapidly changing conditions led to the supplement.

294. National Coal Association. *Coal Data.* 1986 ed. Washington, DC, 1975– . (Annual)

1984–85 statistical data on U.S. coal cover supply, demand, prices, employment, resources, imports and exports. Sources are referenced in this statistical view of world coal.

295. Nyangone, Wellington W. *The O.E.C.D. and Western Multinational Corporations in the Republic of South Africa.* Washington, DC: University Press of America, 1982. 228 p.

Analysis of South African mining, companies which control mining, and mining's effect on apartheid as the public in other countries presses for reduction or termination of activities in the Republic of South Africa. Nyangone is an African specialist in the U.S.

296. Patching, T. H., ed. *Coal in Canada. Canadian Institute of Mining and Metallurgy Special Volume 31.* Montreal, PQ: Canadian Institute of Mining and Metallurgy, 1985. 327 p.

Overall survey of coal mining in Canada, then specifics by district and particular mine.

297. People's Republic of China, Ministry of Coal Industry. *China Coal Industry Yearbook 1982.* First English ed. Hong Kong: Economic Information & Agency, 1983– .

First English publication on coal industry since People's Republic of China began in 1949. Most data are from 1980 and 1981, with highlights from 1949–79 in statistical tables. Mine machinery, including specifications, is described, personnel noted, and monographs on various coal subjects included. An interesting glimpse of a major Chinese industry.

298. Peterson, Jerald R. *Bureau of Mines Research 1986.* Washington, DC: Government Printing Office, 1986. 131 p.

Mining technology research and minerals policy analysis accomplishments for the year are summarized annually. Survey of state-of-the-art in U.S. mining gives a valuable overview of trends in mining.

299. Roberts, Peter W., and Shaw, Tim. *Mineral Resources in Regional and Strategic Planning.* Aldershot, England: Gower Publishing, 1982. 165 p.

U.K. wants to develop own mineral resources logically and lessen dependence on outside sources. Comprehensive, coordinated recommendations for central government policy balancing development need and environmental impact. Thoughtful assessment of a common need in many countries.

300. Roskill Information Services Limited. *The Economics of Gold.* London, 1978– .
Comprehensive survey of gold, its production, uses and trade. Data from published sources are combined and correlated, with much mining information. Major producers are listed, trading partners, production and prices are included in this very useful, but expensive, volume. Updated irregularly. One of a series on economics of commodities, such as silver, platinum and phosphate.

301. Roskill Information Services Limited. *Roskill's Metals Databook.* London, 1979– . (Annual)
Statistics on metal production by metal and company producing the metal worldwide. Historical summary of production, prices and consumption is also provided, plus index of companies listed as leading producers. Company information especially useful, as not easily available for various metals in one source.

302. Themelis, Nickolas J., ed. *Opportunities for a Career in Mining & Metallurgy.* New York: The Mining and Metallurgical Society of America, 1983. 168 p.
Mining, materials science, and metallurgy careers are the focus of this career survey, an overall view of what mining and metallurgical engineers do, followed by information about universities offering relevant courses of study.

303. Thomas, Paul R., and Boyle, Edward H., Jr. *Gold Availability—World.* A Minerals Availability Appraisal. U.S. Bureau of Mines Information Circular 9070. Washington, DC: Government Printing Office, 1986. 87 p.
Primary gold production from 135 mines and deposits worldwide were evaluated for long-term cost and availability. Production from market economy countries is predicted to remain constant, and production may decline significantly after 2010. Useful to assist in planning and financing new gold development.

304. U.S. Bureau of Mines. *Domestic Consumption Trends, 1972–82, and Forecasts to 1993 for 12 Major Metals.* U.S. Bureau of Mines Information Circular 9101. Washington, DC: Government Printing Office, 1986. 34 p.
Trends in intensity of use for major metals, including aluminum, copper, lead and the platinum-group metals, plus forecasts for U.S. use to 1993, according to the Bureau of Mines. Consumption and intensity of use trends were estimated, then forecast to 1993. Only aluminum, platinum-group metals, titanium and tungsten consumption are predicted to increase.

305. U.S. Bureau of Mines. *Mineral Commodity Summaries; An Up-to-Date Summary of 88 Nonfuel Mineral Commodities.* Washington, DC: Government Printing Office, 1987. 189 p.
Brief summary by commodity, with statistics on imports, tariffs and stockpiles. Relatively fast reporting of data, with 1986 data estimated, make this particularly useful to check recent trends.

306. U.S. Bureau of Mines. *Mineral Industries of Africa.* U.S. Bureau of Mines Mineral Perspectives. Washington, DC: Government Printing Office, 1984. 153 p.

Summary of mineral industries in the fifty-two African countries. Reserves, production, trade and major mining companies are covered. Infrastructure is analyzed and future prospects assessed. Useful for basic data and as directory of African mining companies.

307. U.S. Bureau of Mines. *Mineral Industries of Europe and the U.S.S.R.* U.S. Bureau of Mines Mineral Perspectives. Washington, DC: Government Printing Office, 1984. 143 p.

As with the other *Mineral Perspectives,* this covers mining and processing, this time in the twenty-seven countries of Western and Eastern Europe. Production, consumption, trade, and producers of major minerals are summarized.

308. U.S. Bureau of Mines. *Mineral Industries of the Middle East.* U.S. Bureau of Mines Mineral Perspectives. Washington, DC: Government Printing Office, 1986. 66 p.

Analysis of mineral and petroleum industries for the sixteen countries in the Middle East. Supply, trade, and mineral policies are noted.

309. U.S. Geological Survey. *International Strategic Minerals Inventory Summary Report.* U.S. Geological Survey Circular 930. Washington, DC: Government Printing Office, 1984– .

930-A, Manganese, 1984, 22 p.; 930-B, Chromium, 1984, 41 p.; 930-C, Phosphate, 1984, 41 p.; 930-D, Nickel, 1985, 62 p.; 930-E, Platinum-Group Metals, 1986, 34 p. Series of cooperative studies by Australia, Canada, Federal Republic of Germany, Republic of South Africa, United Kingdom and U.S. Each contains overview of the strategic mineral, its uses, distribution, resources and production. Geologic and location information summarized in tables. Useful for current status of selected strategic minerals, with input from many producing countries.

310. Woods, Douglas W., and Burrows, James C. *The World Aluminum-Bauxite Market; Policy Implications for the United States.* A Charles River Associates Research Project. New York: Praeger, 1980. 237 p.

Organization and structure of the world aluminum/bauxite industry and the impacts caused by the International Bauxite Association, the cartel of most non-Communist producers. Supply, market share, demand and projections are presented.

Conferences and Symposia

Selected conferences are recent or ongoing, to cover current technology, supplies, demand and markets in the mineral industries. They are representative of a much larger number of conferences, held around the world, which relate to mining.

311. *Advanced Geostatistics in the Mining Industry.* Proceedings of the NATO Advanced Study Institute, Instituto di Geologia Applicata, University of Rome, Italy, 13–25 October 1975. Edited by M. Guarascii, M. David, and C. Huijbregts. NATO Advanced Study Institutes Series: Series C, Mathematical and Physical Sciences, vol. 24. Hingham, MA: D. Reidel, 1976. 461 p.
Basic concepts of geostatistics, applications in the mining industry and case studies of use. Most papers are in English, a few in French.

312. American Mining Congress. *Session Papers, AMC Mining Convention.* Washington, DC, 1976– . (Annual)
Papers bound by set, on subjects such as technology, environment, public lands, management and economics. The whole spectrum of noncoal mining and mineral industries topics show trends in commodities, locations and technology.

313. American Mining Congress. *Session Papers, Annual Coal Conference.* Washington, DC, 1976– . (Annual)
Session papers published in subject sets, including coal preparation, safety and health management, surface mining and underground mining. No index, just collection of bound prints of papers given.

314. *Application of Rock Mechanics to Cut and Fill Mining.* Proceedings of the Conference on the Application of Rock Mechanics to Cut and Fill Mining, University of Lulea, Sweden, 1–3 June 1980. Edited by Ove Stephansson and Michael J. Jones. London: Institution of Mining and Metallurgy, 1981. 376 p.
State-of-the-art worldwide in the increasing use of rock mechanics in cut-and-fill design, including Scandinavian geomechanics studies and computer models.

315. *Applied Mining Geology.* Symposium Society of Mining Engineers of AIME, Salt Lake City, 1983. Edited by A. J. Erickson, Jr. New York: Society of Mining Engineers of AIME, 1984. 222 p.

Symposium on mining geology, exploration planning, and exploitation of mineral resources to optimize mine profitability. Geology is a partner in the mining venture, to optimize production. Illustrative case studies are presented. Subject index included.

316. *Automation in Mining, Mineral and Metal Processing.* Proceedings of the 3rd IFAC Symposium, Montreal, PQ: 18–20 August 1980. Oxford, New York: Pergamon, 1980. 662 p.

Computer monitoring for safety, equipment, environmental protection, transportation, and metallurgy is the focus. No index.

317. *Broken Hill Conference.* Broken Hill, New South Wales, July 1983 Conference Series. Parkville, Australia: Australasian Institute of Mining and Metallurgy, 1983. 445 p.

Broken Hill, its deposits, mining and impact on Australian mineral industries are discussed in technical, precise papers. Conference location shifts each year, and AIMM publishes all proceedings. Good overview of a different portion of Australian mining each time.

318. *Canadian Rock Mechanics Symposium Proceedings.* Montreal, PQ: Canadian Institute of Mining and Metallurgy, 1962– .

This irregularly held conference covers state-of-the-art in Canadian rock mechanics. The technical papers include many mining applications and case studies.

319. *Coal Resources: Origin, Exploration and Utilization in Australia.* Proceedings of a Symposium of the Coal Group, Geological Society of Australia, Inc, Melbourne, November 15–19, 1982. Edited by C. W. Mallet. Australian Coal Geology, Vol. 4, Parts 1–2. Melbourne, Australia: Geological Society of Australia, 1983. 597 p.

Exploration and mining of Australian low-rank brown coals, including constraints on mine development, are the focus of this symposium volume. Technology and case studies are included.

320. *Congress of the Council of Mining and Metallurgical Institutions, 12th. Proceedings.* Johannesburg, South Africa: South African Institute of Mining and Metallurgy, 1982. 2 vols.

Mineral deposits, mining industry, metallurgical processing, and technical mining in South Africa are emphasized in this conference. Congress location and emphasis changes each time, and publishers vary.

321. *Design and Construction of Tailings Dams.* Proceedings of a Seminar, November 6–7, 1980, Golden, Colorado. Edited by David Wilson. Golden, CO: Colorado School of Mines Press, 1981. 280 p.

Case histories of taconite and oil shale tailings are included in this state-of-the-art seminar on tailings dams. Regulatory aspects of tailings dams are included.

322. *Design and Operation of Caving and Sublevel Stoping Mines.* Proceedings of International Conference on Caving and Sublevel Stoping, Denver, Colorado, 1981. Edited by Daniel R. Stewart. New York: Society of Mining Engineers of AIME, 1981. 843 p.

Practical orientation on current practice in the low-cost underground mining methods of block caving, sublevel caving, and sublevel stoping applications for mining deeper, lower grade mineral deposits. Selection of the appropriate method is addressed, and case studies from around the world are included.

323. *Economics of Mineral Engineering.* An Inter-Regional Seminar Organised by the United Nations in Cooperation with the Government of Turkey, Ankara, April 1976. London: Mining Journal Books, 1976. 223 p.

Mineral industry and international trade, financing mineral developments and metallurgy are covered by scholarly, well-referenced papers.

324. *Finance for the Minerals Industry; Symposium.* New York: Society of Mining Engineers of AIME, 1985. 842 p.

Straightforward, practical view of mine finance, with papers from mining industry personnel and financial experts. The international focus recognizes the nature of mining. Risk analysis, sources of finance, taxation, and case studies of mine finance are covered.

325. *Geology, Mining and Extractive Processing of Uranium.* An International Symposium, London, January 17–19, 1977. Edited by M. J. Jones. London: Institution of Mining and Metallurgy, 1976. 171 p.

Preprints of papers on uranium deposits in Europe and Canada. Reserves, resources and processing are all included, with papers in English, French or Spanish.

326. *Geology of Uranium Deposits.* Proceedings of the CIM-SEG Uranium Symposium, September 1981. Edited by T. I. I. Sigglad and W. Petruck. Canadian Institute of Mining and Metallurgy, Special Volume 32. Montreal, PQ: Canadian Institute of Mining and Metallurgy, 1985. 268 p.

Preprints of papers on uranium deposits in Europe and Canada. Reserves, resources and processing are all included in papers written in English, French or Spanish.

327. *Gold Mining, Metallurgy and Geology.* Regional Conference, Perth and Kalgoorlie Branches, Australasian Institute of Mining and Metallurgy, 9–11 October 1984, Kalgoorlie, Western Australia. Parkville, Australia: Australasian Institute of Mining and Metallurgy, 1985. 477 p.

Economics and ore reserves, gold mining techniques, and geology of gold deposits, primarily in Western Australia, with case studies. An author index is included.

328. *Ground Control in Mining.* Proceedings of the Second Conference on Ground Control in Mining, Morgantown, West Virginia, July 19–21, 1982. Edited by Syd S. Peng. Morgantown, WV: West Virginia University, 1982. 258 p.

Conference covered longwall mining, roof control and monitoring, mine design and surface subsidence. Papers are technical, oriented toward practicing engineers.

329. *Ground Control in Room-and-Pillar Mining.* Conference, Southern Illinois University—Carbondale, August 6–8, 1980. Edited by P. Yoginder Chugh. New York: Society of Mining Engineers of AIME, 1982. 157 p.

Ground control in U.S. coal and noncoal mines, plus Polish copper mines. Papers discuss prevention of roof, rib and sides fall, plus surface subsidence.

330. *IAGOD Symposium.* Proceedings of the Fifth Quadrennial. Edited by John Drew Ridge. Stuttgart, West Germany: Schweizerbart'sche Verlagsbuchhandlung, 1980. 2 vols.

Studies on ore deposits and regions are covered in the symposia of the International Association on the Genesis of Ore Deposits. The scholarly papers are written in English and contain many citations to further sources of information. Each symposium increases knowledge of ore deposits worldwide, and each is edited and published separately .

331. *Industrial Minerals International Congress; Proceedings.* London: Metal Bulletin Books Ltd., 1974– . (Semiannual)

Proceedings of semiannual conferences on industrial rocks and minerals worldwide. Useful for surveys of industrial minerals and their markets in a particular country or worldwide. Specific deposits are also covered in this industry-oriented series.

332. *Interfacing Technologies in Solution Mining.* Proceedings of the Second SME—SPE International Solution Mining Symposium, Denver, Colorado, November 18–20, 1981. Edited by W. J. Schlitt. New York: American Institute of Mining, Metallurgical and Petroleum Engineers, 1982. 370 p.

Solution mining for trona, salt, gold, silver and uranium receive international coverage with one-third of the papers from outside the United States. Many case studies in the technical presentations.

333. *International Conference on the Hydraulic Transport of Solids in Pipes.* Papers Presented at Hydrotransport. Cranfield, England: BHRA Engineering, 1970– . (Irregular)

Slurry transport of different ores and mineral wastes, such as coal, iron, bauxite, gold. Fluid mechanics aspects are emphasized and many case studies presented. *Hydrotransport* is the basic source of current information on slurry technology and practice worldwide.

334. *International Iron Ore Symposium.* First, Amsterdam, The Netherlands, March 26–27, 1979; Proceedings. Edited by Raymond Cordero. Surrey, England: Metal Bulletin Ltd., 1979. 184 p.
Iron industry symposium focuses on trends in supply, shipping, and iron ore consumption. Steel consumption was already dropping, and the international participants concentrated on economics and trends in the iron industry.

335. *International Minerals; A National Perspective.* Symposium, AAAS Annual Meeting, Washington, DC, January 3–8, 1982. Edited by Allen Agnew. Boulder, CO: Westview Press, 1983. 164 p.
World mineral resources are examined in regard to U.S. production and self-sufficiency. Mineral deposits and development in areas such as Alaska, South Africa and U.S.S.R. are discussed, and U.S. policies summarized.

336. *International Symposium on the Applications of Computers and Operations Research in the Mineral Industries.* Brisbane, Australia: July 4–8, 1977; APCOM 77. Parkville, Australia, Australasian Institute of Mining and Metallurgy, 1977. 520 p.
Operations research in Australia, Sweden, Germany, plus computer models for operations management and mine planning. Models also include resource planning and investment used in mine situations and results.

337. The Israeli Association for the Advancement of Mineral Engineering. *The Eighth Conference.* Channuka, Israel, December 29–30, 1986. Jerusalem: Geological Survey of Israel, 1987. 96 p. in English, 242 p. in Hebrew.
Mineral deposits and mining in Israel, including use of microcomputers in phosphate mining, bulk handling, and haulage road design. The conference was sponsored by the Ministry of Commerce & Industry, the Ministry of Energy & Infrastructure, and Mining Industries.

338. *Large Open Pit Mining Conference.* Newman, Western Australia, October 1986. Edited by J. R. Davidson. Parkville, Australia: Australasian Institute of Mining and Metallurgy, 1986. 297 p.
State-of-the-art in Australian large open-pit mining, with feasibility studies, uses of geostatistics, and design factors of large mines.

339. *Mine Ventilation.* Proceedings of the 2d U.S. Mine Ventilation Symposium, University of Nevada—Reno, September 23–25, 1985. Edited by Pierre Mousset Jones. Rotterdam, The Netherlands; Boston: Balkema, 1985. 780 p.
Mine ventilation topics at the core of this symposium are booster fans and use of recirculated air, plus ventilation necessary for future storage of nuclear waste, methane control in coal mines, and microcomputer modeling of mine ventilation.

340. *Mineral Waste Utilization Symposium.* Chicago: IIT Research, 1971– . (Annual)
Mining wastes and their uses, recovery from tailings, use in construction, and scrap-metal applications are covered in these symposia. The broad spectrum of uses for mineral wastes are presented, and current research results reported.

341. *Mining Techniques, Mining Equipment.* Proceedings of the Second NMIMT Symposium, Socorro, New Mexico, April 21–22, 1983. Socorro, NM: New Mexico Institute of Mines and Technology, 1983. 5 vols.
State-of-the-art survey of surface mining, shaft sinking, underground mining, and bucket-wheel excavators in the southwestern U.S. Sponsored by NMIMT, Cooney Mining Club and New Mexico Mining Association.

342. *The Occurrence, Prediction and Control of Outbursts in Coal Mines.* Symposium, Southern Queensland Branch, Australasian Institute of Mining and Metallurgy, September 1980. Parkville, Australia: Australasian Institute of Mining and Metallurgy, 1980. 239 p.
Primarily Australian experience and case studies of rock bursts in underground coal mines. Causes are analyzed and cures suggested.

343. *Precious Metals in the Northern Cordillera.* Proceedings of a Symposium, Vancouver, British Columbia, Canada, April 13–15, 1981. Edited by A. A. Levinson. Association of Exploration Geochemists Special Publication Number 10. Rexdale, ON: Association of Exploration Geochemists, 1982. 214 p.
Geochemical methods and case histories on exploration for gold and silver illustrate analytical and exploration techniques, plus characteristics of ore deposits. Conference was held jointly with The Cordilleran Section of the Geological Association of Canada.

344. *Project Development.* Second Symposium Australasian Institute of Mining and Metallurgy, Sydney, October 1986. Parkville, Australia: Australasian Institute of Mining and Metallurgy, 1986. 324 p.
The mining project, papers on risk assessment and working capital, feasiblity studies, and case studies of actual mine commissioning are covered in this unusually titled symposium.

345. *Project Development.* Symposium, Australasian Institute of Mining and Metallurgy, Sydney Branch, Sydney, New South Wales, Australia, November 14–16, 1983. Parkville, Australia: AIMM, 1983. 416 p.
Symposium on the facets of Australian mine development, from project evaluation to marketing, regulations, finance and industrial relations. Specific case studies are included, and an author index is provided.

346. *Research and Engineering Applications in Rock Masses.* Proceedings of the 26th U.S. Symposium on Rock Mechanics, South Dakota School of Mines and Technology, Rapid City, June 26–28, 1985. Edited by Eileen Ashworth. Rotterdam, The Netherlands; Boston: Balkema. 2 vols. 2,292 p.

The U.S. rock mechanics symposia are published by various associations and institutions, but a list of the previous ones is included in this volume. Much mining application is included. Basic series for engineering aspects of mining.

347. *Rockbursts: Prediction and Control.* Symposium, London, October 20, 1983. London: Institution of Mining and Metallurgy, 1983. 173 p.
Symposium on occurrence, prediction, control and monitoring of rock bursts, primarily in mines, their prediction and control. Case histories from around the world are presented, and safety aspects of rockburst prevention included.

348. Rocky Mountain Mineral Law Institute. *Proceedings of the Annual Institute.* New York: Matthew Bender, 1955– . (Annual)
Lawyers discuss mineral ownership, contracts, mining, milling, water and reclamation in well-referenced legal opinions which are also comprehensible, often interesting, and always informative. Topics generally timely and relevant to current conditions in the industry.

349. *Safety in Mining Research.* Proceedings of the 21st International Conference of Safety in Mines Research, Sydney, Australia, October 21–25, 1985. Edited by A. R. Green. Rotterdam, The Netherlands; Boston: Balkema, 1985. 782 p.
Very technical presentations on strata control, rock bursts, fires and explosions by researchers around the world.

350. *Selective Open Pit Gold Mining.* Seminar, Australasian Institute of Mining and Metallurgy, Perth Branch, Perth, May 1986. Parkville, Australia: Australasian Institute of Mining and Metallurgy, 1986. 169 p.
Practical reporting of gold mines in Australia, from reserve estimation through sampling to selective mining as currently practiced in low-grade deposits.

351. Society of Mining Engineers of AIME. *Preprints, Annual Conference.* Littleton, CO, 1959?– . (Annual)
Preprints from the annual SME conference are available as separates, which cover worldwide mining and processing but emphasize North America. Managers and engineers present papers on financial and marketing aspects, new technology, case studies and projections. Some papers will later be published in *Mining Engineering,* but most are available only as preprints and can be ordered separately.

352. *Solution Mining Symposium 1974.* Proceedings of a Symposium 103rd AIME Annual Meeting, Dallas, Texas, February 25–27, 1974. Edited by F. F. Aplan, W. A. McKinney, and A. D. Pernichele. New York: American Institute of Mining, Metallurgical and Petroleum Engineers, 1974. 467 p.
Because solution mining often offers cost reductions over other mining, this symposium examines rock fracturing, fluid flow, constraints, and practical applications of solution mining for gold, copper, and uranium.

353. *Stability in Underground Mining.* First International Conference, Vancouver, British Columbia, Canada, August 16–18, 1982. Edited by C. O. Brawner. New York: Society of Mining Engineers of AIME, 1983. 1,071 p.
Review of stability problems in metal and nonmetal mines, including costs, safety and productivity. Practical reports of stability problems and solutions in coal, nickel and oil-shale mines are presented. Significant international coverage.

354. *Strategic Minerals: A Resource Crisis.* Conference, Council on Economics and National Security, May 22, 1981. New York, 1981. 105 p.
Overdependence of U.S. on foreign sources of strategic minerals such as manganese, chromium, cobalt and platinum is examined. Conference speakers press for less dependence and more U.S. production, so national defense remains strong.

355. *Symposium on Oil Shale.* Proceedings. Golden, CO: School of Mines Press, 1964– . (Annual)
State-of-the-art in oil shale, its deposits, mining, technology, processing, waste disposal, environmental aspects and financial feasibility. The symposia were published in the *Colorado School of Mines Quarterly,* 1964–76, and then as separate publications.

356. *Symposium on Salt.* Proceedings. Alexandria, VA: The Salt Institute; Cleveland, OH: Northern Ohio Geological Society, 1964– . (Annual)
Dry and solution mining are major topics in each volume, and rock mechanics, underground storage and geology are also covered. Other evaporites are considered, in addition to salt. The first five symposia proceedings were published by the Northern Ohio Geological Society.

357. *Titanium '80; Science and Technology.* Proceedings of the Fourth International Conference on Titanium, Kyoto, Japan, May 19–22, 1980. Edited by H. Kimura and O. Izumi. New York: American Institute of Mining, Metallurgical and Petroleum Engineers, 1980. 4 vols.
Overviews of titanium industry in various countries are presented, along with substantial information on titanium and its alloys, their manufacture and uses. A comprehensive picture of titanium.

358. *Tungsten: 1982.* Proceedings of the Second International Tungsten Symposium, San Francisco, June 1–5, 1982. London: Mining Journal, 1982. 179 p.
All aspects of tungsten, from chemistry to finance and trade, focus on the tungsten market and its economics in this worldwide conference.

359. U.S. Bureau of Mines. *Control of Acid Mine Drainage.* Proceedings of a Technology Transfer Seminar, Pittsburgh, PA, April 3–4, 1985. U.S. Bureau of Mines Information Circular 9027, Washington, DC: Government Printing Office, 1985. 61 p.

One of a continuing series of Technology Transfer Seminars, this one describes new methods for treating acid mine water. Water treatment, reaction retardation and preventitive actions are recommended, based on Bureau of Mines research. Others can be accessed through the Bureau of Mines indexes.

360. U.S. Bureau of Mines. *Precious Metals Recovery from Low Grade Resources.* Proceedings, Bureau of Mines Open Industry Briefing Session at the National Western Mining Conference, Denver, Colorado, February 12, 1986. U.S. Bureau of Mines Information Circular 9059. Washington, DC: Government Printing Office, 1986. 56 p.

One of a continuing series of Open Industry Briefing Sessions, this highlighted recent Bureau of Mines research, some ongoing, to recover gold and other precious metals from low-grade resources. Heap leaching, recovery from scrap, and cyanidation of gold ores are analyzed.

361. *U.S. Water Jet Symposium.* First, Golden, Colorado, April 7–9, 1981, Proceedings. Edited by Fun-Den Wang, Levant Ozdemir, and Russell J. Miller. Golden, CO: Colorado School of Mines Press, 1982. 248 p.

Reports on technology and research in progress relate to the mining industry's use of water jets for development and drilling in uranium and coal. The conference focused on mining but also covered other industrial uses of water jets.

362. *Ventilation of Coal Mines.* Proceedings of the Symposium Australasian Institute of Mining and Metallurgy, Illawarra Branch, Wollogong Technical College, Robertson, New South Wales, May 10–13, 1983. Robertson, Australia: Australasian Institute of Mining and Metallurgy, 1983. 1 vol. (Various paging)

Mostly Australian examples of coal mine ventilation, but some discussions about U.K. and Germany. Various facets of coal mine ventilation are covered.

Texts and Treatises

Basic texts on mining and mineral industries are necessary for a mining collection. Business and technological treatises are arranged by general subject.

INVESTMENTS

Investment guides are geared toward the novice or the experienced trader, for actual mineral commodities, futures or company stocks. In addition to sources listed here, a primary source of mining company information is available through its annual report, which can be requested directly from the company, using addresses in the various directories.

363. *The Brokers' Guide to North American Resource Companies.* Portland, OR: Brokers' Guide, 1984– . (Unpaged)
Annually updated guide with two-page annual report type of information on U.S. and Canadian mining companies. Useful for reserves, acquisitions, and basic financial data. More information can be requested through Brokers' Guide.

364. Cavelti, Peter C. *New Profits in Gold, Silver, Strategic Metals; The Complete Investment Guide.* New York: McGraw-Hill, 1985. 221 p.
Written for the layman, guide presents advantages and disadvantages of investing in gold, silver and strategic metals. Storage and insurance of the commodity, accounts so the investor does not deal in the actual metals, and gold futures are options assessed.

365. Dammert, Alfredo, and Palaniappan, Sethu. *Modelling Investments in the World Copper Sector.* Austin, TX: University of Texas Press, 1985. 136 p.
Computer modeling mining, processing and fabrication of copper worldwide. Both developed and developing countries are concerned with copper industry, costs, capacity, investment and markets. Forecasts to 2000 are included.

366. Financial Post. *Survey of Mines and Energy Resources.* Toronto, ON: Maclean Hunter, 1982. 540 p.

Canadian mining and energy companies are presented with the price range of their stocks, and two years of financial statements. Activity, reserves and production are summarized for most companies. Varied amounts of information are included, some are quite brief but most are substantial. Very useful for comparing Canadian mining companies.

367. Gibson-Jarvie, Robert. *The London Metal Exchange; A Commodity Market.* 2d ed. New York: Nichols Publishing, 1983. 203 p.

The London Metal Exchange's history, philosophy, and contract process are assessed; then particular metals and their trading are analyzed. Describes how commodity trading really works. A glossary of trading terms and listing of LME registered brands, plus an example LME contract. Detailed index.

368. *Guide to World Commodity Markets.* Edited by Ethel De Kloper, New York: Nichols Publishing, 1977. 308 p.

Commodity markets, including copper, lead, silver, tin and zinc, are profiled for location, rules, and arbitration. The role and function of world commodity markets are covered, and traded commodities are surveyed for the preceeding five years. Important for understanding the process of commodities trading and the locations where trading occurs.

369. Price Waterhouse. *Financing Foreign Development of Non-Fuel Mineral Resources; An Analysis of Loan Terms of International Financial Institutions vs. Those of Commercial Banks.* U.S. Bureau of Mines Open-File Report 52-86. Pittsburgh, PA: U.S. Bureau of Mines, 1986. 310 p. PB 86-247228.

Commercial loan terms for seven mineral projects in developing countries were compared, and costs of financing analyzed in relation to costs of long-term debt of five large U.S. minerals producers. Timely review because of increased foreign competition impacts on U.S. mining companies.

370. Sinclair, James E., and Parker, Robert. *The Strategic Metals War; The Current Crisis and Your Investment Opportunities.* New York: Arlington House, 1983. 185 p.

Strategic metals, their occurrence, possible supply disruptions, and investments in the metals or shares in metals companies are explained in how the market works. Investment in the strategic metals is recommended and reasons presented.

EXPLORATION AND RESOURCES

Exploration for orebodies and resource evaluations are the basis of all mining projects and mineral industries. These basic texts provide the foundation to lead to further information on mineral deposits and districts.

371. American Institute of Mining and Metallurgical Engineers. *Ore Deposits of the Western States.* Lindgren Vol., 1st ed. New York, 1933. 797 p.
Very useful, this classic work classifies ore deposits and describes deposits in the western United States. Index by deposit, type and location.

372. Averitt, Paul. *Coal Resources of the United States.* U.S. Geological Survey Bulletin 1412. Washington, DC: Government Printing Office, 1974. 131 p.
Quantity and distribution of U.S. coal resources are estimated, including coking coal and strippable coal as separate categories. Production is summarized and future uses projected. Extensive reference list and detailed index.

373. Brobst, Donald A., and Walden, P., eds. *United States Mineral Resources.* U.S. Geological Survey Professional Paper 820. Washington, DC: Government Printing Office, 1973. 722 p.
Uses, technology and economics of commodities are analyzed, along with geology of the deposits in this excellent summary of U.S. mineral resources and deposits.

374. Coppa, Luis V., and DiFrancesco, Carl A. *Minerals Availability Directory of Mineral Deposits for 23 Strategic Mineral Commodities.* U.S. Bureau of Mines Open-File Report 152-84. Pittsburgh, PA: U.S. Bureau of Mines, 1984. 216 p. PB 84-239185.
Directory of twenty-three major mineral commodities, both in the U.S. and worldwide. Location, commodities, company, and mining method are assessed in current study.

375. Cronan, D. S. *Underwater Minerals.* London, New York: Academic Press, 1980. 362 p.
Geology and location of minerals, with emphasis on manganese nodules, but metalliferous sediments also considered. The various types of underwater minerals are viewed as a group and mining methods presented.

376. De Geoffroy, J. G., and Wignall, T. K. *Designing Optimal Strategies for Mineral Exploration.* New York, London: Plenum, 1985. 364 p.
Efficient use of technology to explore for hard-rock ore deposits is examined in relation to current budget constraints in the mineral industries. Survey designs for mineral detection and optimizing drill testing found by survey are covered. Extensive, specific bibliographies lead to more information.

377. Elevatorski, E. A. *Gold Mines of the World.* Dana Point, CA: Minobras, 1981. 107 p.
Gold resources and mining are summarized, and some mines are mentioned by U.S., state and foreign countries. Overall view is useful introduction, and then more detailed sources can be consulted. No references or index.

378. Elevatorski, E. A. *Molybdenum Resources Guidebook.* Dana Point, CA: Minobras, 1980. 159 p.
Overall summary of molybdenum deposits and mines, first in the U.S., then the rest of the world. Sketch maps with deposit locations are included. References are listed.

379. Evans, E. L. *Uranium Deposits of Canada.* Canadian Institute of Mining and Metallurgy Special Vol. 33. Montreal, PQ: Canadian Institute of Mining and Metallurgy, 1986. 324 p.
All types of uranium deposits in Canada are covered and mines described. Deposits are grouped by type. Addresses of authors are included to facilitate access to further information.

380. Gaschnig, John. *Development of Uranium Exploration Models for the Prospector Consultant System.* Menlo Park, CA: SRI International, 1980. 603 p.
Three uranium exploration models were developed for the computer program Prospector. Geologists' inputs were programmed as decision points to approximate artificial intelligence. Regional and prospect models were then validated through comparison with geologists' conclusions.

381. Habashi, F., and Bassyouni, F. A. *Mineral Resources of the Arab Countries.* 2d ed. London: Chemecon Publishing Limited, 1982. 60 p.
Metallic and nonmetallic mineral resources of each Arab country are briefly profiled from published reports, conference papers, and government publications. Colored maps of mineral locations in each country and by commodity throughout the region.

382. *Handbook of Strata-Bound and Stratiform Ore Deposits.* Amsterdam, The Netherlands: Elsevier, 1976–81. Vols. 1–7, 1976; Vols. 8–10, 1981.
Comprehensive scholarly treatise includes characteristics of the ores, regional studies of deposits such as nickel, iron, copper, tin, lead and zinc, plus sections on specific deposits. Extensive bibliographies and a wealth of well-researched information.

383. Ishihara, Shunso, ed. *Geology of Kuroko Deposits. Mining Geology* Special Issue No. 6. Tokyo, Japan: The Society of Mining Geologists of Japan, 1974. 435 p.
Papers from the journal, *Mining Geology,* revised, updated and translated into English, if necessary, on Kuroko deposits, with information from more than twenty mines in Japan. Author, subject, and locality indexes.

384. Ishihara, Shunso and Takenouchi, Sukune, eds. *Granitic Magmatism and Related Mineralization. Mining Geology* Special Issue No. 8. Tokyo, Japan: The Society of Mining Geologists of Japan, 1980. 247 p.
Ore deposits and districts in Japan, China and Southeast Asia are profiled in papers written in English. Useful for insight into areas not often included in mineral assessments.

385. Jensen, Mead L., and Bateman, Alan M. *Economic Mineral Deposits.* 3d ed. New York: Wiley, 1979. 593 p.
Basic text describes metallic and nonmetallic mineral deposits.

386. Kent, Percy Edward, Sir. *Minerals from the Marine Environment.* Resource and Environmental Sciences Series. New York: Wiley, 1980. 88 p.
Readable overview of deep-sea minerals, beach sands, shallow sea minerals and minerals from solution. Probable recovery, economics, and the environmental problems and consequences of mining deep-sea minerals are assessed.

387. Knight, C. L., ed. *Economic Geology of Australia and Papua New Guinea.* Australasian Institute of Mining and Metallurgy Monographs 5–8. Parkville, Australia: Australasian Institute of Mining and Metallurgy, 1975–76. 4 vols.
The four volumes cover metals, coal, petroleum, industrial minerals and rocks comprehensively. Australian and New Guinean mineral deposits are described, along with their characteristics, ages, and locations. Sketch maps included.

388. Koschmann, A. H., and Bergendahl, M. H. *Principal Gold-Producing Districts of the United States.* U.S. Geological Survey Professional Paper 610. Washington, DC: Government Printing Office, 1968. 283 p.
Geology, mining history and production of the major gold districts in the western U.S., plus Tennessee and North Carolina. Index by mine and district. Very useful for data on U.S. gold mines and mining districts.

389. Merritt, Roy D. *Coal Exploration, Mine Planning, and Development.* Park Ridge, NJ: Noyes Publications, 1986. 464 p.
Overview of coal geology, exploration, mining and technology is geared to those inside mining industry. Detailed subject index and ninety-four-page glossary, with references, complete the volume.

390. Panczner, William D. *Minerals of Mexico.* New York: Van Nostrand Reinhold, 1987. 459 p.
Mexico's mining history and mining districts are summarized; also given are an alphabetical listing of minerals, their descriptions and locations.

391. Park, Charles F., Jr., and MacDiarmid, Roy A. *Ore Deposits.* 3d ed. San Francisco, CA: Miller-Freeman, 1975. 530 p.
The basic text on ore deposits, from deposition to types of deposits, such as pegmatites, sedimentary, epithermal. Case studies of specific deposits included. Newest edition, *Geology of Ore Deposits* is entry 19 in the Core Library Collection.

392. Ramdour, Paul. *The Ore Minerals and Their Intergrowths.* 2d ed. English translation of the 4th ed. New York: Pergamon, 1980. 2 vols.
Scholarly, comprehensive treatise on ore minerals. Genetic systematics of ore deposits, principles of classification, and descriptions of ore minerals are detailed in this basic text.

393. Rich, Robert A.; Holland, Heinrich D.; and Peterson, Ulrich. *Hydrothermal Uranium Deposits.* Developments in Economic Geology, 6. Amsterdam, The Netherlands; Oxford, New York: Elsevier, 1977. 264 p.
Scholarly overview of hydrothermal uranium deposits plus descriptions of specific deposits throughout the world.

394. Ridge, John D., ed. *Ore Deposits of the United States, 1933–1967; The Graton-Sales Volume.* New York: American Institute of Mining, Metallurgical and Petroleum Engineers, 1968. 2 vols. 1,880 p.
Survey of the major ore deposit districts. A good place to start research on classic U.S. mineral deposits, even though dated. Much mining information is also covered.

395. Scalisi, Philip, and Cook, David. *Classic Mineral Localities of the World, Asia and Australia.* New York: Van Nostrand Reinhold, 1983. 226 p.
Exceptional mineral quality specimens are the focus of this study, with mine districts and single mines profiled for history and mineral specimens. Minerals are described in detail, with drawings of crystal forms and photographs of actual specimens.

396. Sinkankas, John. *Prospecting for Gemstones and Minerals.* New York: Van Nostrand Reinhold, 1970. 397 p.
Basic text for amateur prospectors advises the beginner on equipment, features of minerals in the field, how to collect specimens and how to prepare, store and exhibit them.

397. Smirnov, V. I., ed. *Ore Deposits of the U.S.S.R.* Translated by D. A. Brown. Belmont, CA: Pitman Publishing, 1977. 2 vols.
Group characteristics of ore deposits, discussed by commodity, such as iron. Specific deposits in the USSR are described in detail.

398. Stanton, R. L. *Ore Petrology.* New York: McGraw-Hill, 1972. 713 p.
Basic text on scientific investigation for ore deposits. Genetic theories evenly presented for all types of ores and origins. Personal name and subject index, plus extensive bibliography.

399. Titley, Spencer R. *Advances in Geology of the Porphyry Copper Deposits, Southwestern U.S.* Tucson: AZ: University of Arizona Press, 1982. 560 p.
Copper deposits are covered for which published information was not yet available. Twelve specific copper deposits and one copper mining district in the southwestern U.S., primarily Arizona, are profiled. A comprehensive index aids in research.

400. *Vein Type Uranium Deposits; Report of the Working Group on Uranium Geology Organized by the International Atomic Energy Agency.* Vienna, Austria: International Atomic Energy Agency, 1986. 423 p. IAEA-TECDOC-361.
Vein-type uranium deposits, their origin and description, are written by expert members of the working group, who view vein-type deposits as a significant resource for future development. Deposits in India, Sweden, Australia and other countries are profiled, and most entries are in English.

401. Ward, Colin R., ed. *Coal Geology and Coal Technology.* London; Palo Alto, CA: Blackwell Scientific Publications, 1984. 345 p.
Exploration, mining, processing and use of coal deposits are summarized in this current overview of the worldwide coal industry.

402. Wolfe, John A. *Mineral Resources; A World Review.* New York: Chapman and Hall, 1984. 293 p.
Written for the person interested but not experienced in mineral resources by an author who believes in mining. Mineral policies are surveyed; reserves and exploration are explained in a chatty fashion. History of metals and non-metals is covered; current status and outlook are included.

403. *World Uranium Geology and Resource Potential.* The Joint Steering Group on Uranium Resources of the OECD Nuclear Energy Agency and the International Atomic Energy Agency. San Francisco, CA: Miller-Freeman, 1980. 524 p.
Uranium resources in 185 countries are covered—geology, exploration, occurrences and past production, plus potential for new discoveries. Excellent overview of world uranium resources, arranged by continent and then by country. Index is by country.

TECHNOLOGY AND MINING

404. Agricola, Georguis. *De Re Metallica.* Translated by Herbert Clark Hoover and Lou Henry Hoover. New York: Dover, 1950. 638 p.
First published in 1556, this true classic traces the development of mining, metallurgy, geology, mineralogy and mining law. Essential for perspective on very early mining.

405. Biron, Cemal, and Arioglu, Ergin. *Design of Supports in Mines.* New York: Wiley, 1983. 248 p.
Technical study of roof supports includes wooden, steel, bolts, and concrete. Many diagrams of supports and formulas for calculations. References and index are included.

406. Britton, Scott G. *Construction Engineering in Underground Coal Mines.* New York: Society of Mining Engineers of AIME, 1983. 312 p.
All information necessary for underground construction, including equipment specifications, equipment manufacturers, scheduling, costs and wages, is gathered in one volume of practical instruction.

407. Chernigovskii, A. A. *Application of Directional Blasting in Mining and Civil Engineering.* 2d ed., revised and enlarged. Translated from Russian. New Delhi, India: Amerind Publishing Co. Pvt. Ltd., 1985. 318 p.

Practical recommendations for directional blasting and criteria for cost effectiveness are reported, based on Russian experience in mining. Audience is engineers and technicians planning blasting. Ballistics tables are included.

408. Chang, Jui-Lin. *A Study on the Selection of Overburden Handling Systems and Environmental Protection in Surface Mining.* Unpublished Ph.D. dissertation, West Virginia University, 1985. 180 p.

Seven stripping techniques and seventeen combinations of layout and overburdened handling systems are analyzed, equipment investigated and costs assessed in order to select the optimum combination of efficiency and environmental protection.

409. Clement, Wallace. *Hardrock Mining; Industrial Relations and Technological Changes at INCO.* Toronto, ON: McClelland and Stewart Ltd., 1981. 392 p.

Through a case study of Canadian mining at Inco's Sudbury nickel mine, class relationships in mining are explored. Many quotes from miners are included, and some from management and engineers, to give a picture of actual mining conditions. Index included.

410. Crawford, John T., III, and Hustrulid, William A., eds. *Open Pit Mine Planning and Design.* New York: American Institute of Mining, Metallurgical and Petroleum Engineers, 1979. 367 p.

Engineering principles and rules of thumb for open pit mining are contributed mostly by practicing mining engineers and mining consultants. The volume results from a workshop held at the 1978 AIME annual meeting. Very detailed index.

411. Curth, Ernest A., and Listak, Jeffrey M. *Longwall Roof Support Technology in the Eighties: A State-of-the-Art Report.* U.S. Bureau of Mines Information Circular 9066. Washington, DC: Government Printing Office, 1986. 43 p.

Shield faces, the current technology, are assessed, load predicted, and factors contributing to ground control are evaluated. Data are designed to help in roof support selection to continue safety improvement in mines.

412. Douglas, William J., ed. *Production Systems Analysis in Underground Coal Mines.* New York: McGraw-Hill, 1980. 369 p.

An industrial engineering approach to evaluate underground mining systems through analytical methods, modeling and performance measurements. Systematic analysis of mining components and events to improve efficiency in underground coal mines.

413. Farmer, Ian. *Coal Mine Structures.* London; New York: Chapman and Hall, 1985. 310 p.

Results of a six-year study in England on behavior of underground openings in British coal mines. Laboratory and field data show how underground structures related to deformation characteristics of surrounding rocks. Very technical assessment of deformation and failure in coal mines.

414. Green, Robert E., and Ottley, Derek J. *Development of Engineering and Cost Data for Foreign Gold and Silver Properties.* U.S. Bureau of Mines Open-File Report 140-84. Pittsburgh, PA: U.S. Bureau of Mines, 1983. 49 p.

Over 100 technical reports on major gold and silver mines and deposits of the world, outside the United States, are evaluated. Reserves and resources, assays, production, and operating costs are given in ranges to protect confidentiality of mining companies. Significant cost data for comparison with other prospects.

415. Hall, C. J. *Mine Ventilation Engineering.* New York: Society of Mining Engineers of AIME, 1981. 344 p.

Written for the practicing mine ventilation engineer and university students, this text contains many formulas which presume a substantial mathematical background. Natural ventilation, fans, and instrumentation are all assessed.

416. Hartman, Howard L., ed. *Mine Ventilation and Air Conditioning.* 2d ed. New York: Wiley, 1982. 791 p.

Control of the total mine environment is planned, the economics of ventilation and the relative merit of equipment types are assessed. Operating and maintenance principles are illustrated by many formulas, and cooling loads are calculated. Computer applications, plus tunnel ventilation, fires and explosions are covered.

417. Hatkoff, Reed A. *The Encyclopedia of Placer Mining.* Denver, CO: Inter Resources, 1983. 512 p.

Although named an encyclopedia, this volume consists of reprints of U.S. Bureau of Mines Information Circulars 8462, 8517, and 6786, plus other direct reprints from earlier publications. Dates of the publications range from the 1930s to the 1970s. No index or table of contents is included. Much of the early information on placer mining is still valid, however, because the technology can still be used, especially for small-scale mining.

418. Humphreys, Kenneth K., and Leonard, Joseph W. *Basic Mathematics and Computer Techniques for Coal Preparation and Mining.* Energy, Power and Environment Series. New York: Marcel Dekker, 1983. 233 p.

Developed by the Coal Bureau at West Virginia University, practical operating decision techniques comprise the first part of the study, and statistical analysis and computer programs comprise the remainder. FORTRAN programming experience is needed.

419. Krantz, Gary W. *Selected Pneumatic Gunites for Use in Underground Mining: A Comparative Engineering Analysis.* U.S. Bureau of Mines Information Circular 8984. Washington, DC: Government Printing Office, 1984. 64 p.

Fibered, portland-cement based gunite products were tested for sealing, spill prevention, or roof stability control. Practical evaluation of gunite use for underground mining.

420. Kratzsch, Helmut. *Mining Subsidence Engineering.* Translated by R. F. S. Fleming. New York: Springer-Verlag, 1983. 543 p.

Translation of technical German text, revised and enlarged. From mine excavation to subsidence damage, mine subsidence is examined. Sections on surface damage are planned for planners and landowners as well as engineers.

421. Langefors, U., and Kihlstrom, S. *The Modern Technique of Rock Blasting.* 2d ed. Stockholm, Sweden: Almquist & Wiksell, 1963. 405 p.

Rock blasting in mining and tunneling is examined, with many Swedish examples based on practical experience. Methods to calculate the proper blasting charge are presented.

422. Lonergan, James Edward. *Computerized Solutions to Mine Planning and Blending Problems.* Unpublished Ph.D. dissertation, University of Arizona, 1983. 130 p.

Mine planning is proposed to predict sulfur level of coal shipments and allow blending the optimum combination to fulfill coal contracts. The developed interactive blending program was used at one mine, and results were satisfactory, validating the computerized program.

423. Morley, Lloyd A. *Mine Power Systems.* U.S. Bureau of Mines Open-File Reports 178(1)-82, 178(2)-82. vols. I–II. Pittsburgh, PA: U.S. Bureau of Mines, 1982. PB 83-120378, 534 p.; PB 83-120386, 568 p.

Coal mine electrical power systems are assessed comprehensively, with all considerations for planning and designing the proper system. Equipment is covered in Volume II, along with maintenance. Hazards are investigated and recommendations for amelioration included.

424. Ofengenden, N. E., and Kzhvarsheishvili, A. G. *Technology of Hydromining and Hydrotransport of Coal.* Translated by Albert L. Peabody. Rockville, MD: Terraspace, Inc. 1981. 251 p.

Methods of hydrotransport of coal are illustrated by hydraulic mining examples and experience from U.S.S.R. Technical publication is written for engineers and technical workers.

425. Olaf, Jorn H. E. *Automation and Remote Monitoring and Control in Mines.* Translated from German by Bruno E. Sabels and Humphrey G. Cook. Essen, West Germany: Verlag Gluckauf GMBH, 1979. 186 p.

Summary of monitoring and control concentrates on desirable characteristics, not item-specific recommendations. Systems focus on novel techniques, and technology and interdependence of systems are emphasized.

426. Ramani, R. V., ed. *Longwall-Shortwall Mining, State of the Art.* New York: Society of Mining Engineers of AIME, 1981. 296 p.
Longwall and shortwall mining equipment, environment, unit operations, and case studies are included in this volume based on conference papers from 1981, plus selected additions.

427. Stefanko, Robert. *Coal Mining Technology: Theory and Practice.* New York: Society of Mining Engineers of AIME, 1983. 410 p.
A textbook on underground and surface mining considers mine planning, ventilation, ground control, drainage, communication and lighting for an overall view. Practical examples for calculations are included.

428. Tarjan, Gusztav. *Mineral Processing; Comminution, Sizing and Classification,* vol. 1. Budapest, Hungary: Akademiai Kiado, 1981. 583 p.
Preparation, such as crushing and grinding, accompany metallurgical principles in this basic explanation. Charts and tables complete the technical assessment, and a subject index aids in use.

429. Trotter, Donald A. *The Lighting of Underground Mines.* Reprint, 1st ed. Trans Tech Series on Mining Engineering, vol 2. Houston, TX: Gulf Publishing, 1983. 201 p.
A reference and design guide geared to mining engineers who may well not have lighting knowledge. A concern for the human environment is stated, so safety is carefully considered. Standards and guidelines for design of lighting systems are included. A glossary of electrical terms and a detailed index complete the volume.

430. University of Utah Research Institute. *In Situ Leaching and Solution Mining: Evaluation of the State-of-the-Art.* U.S. Bureau of Mines Open-File Reports 64(1-6)-85. Pittsburgh, PA: U.S. Bureau of Mines, 1983. 6 vols.
A thorough state-of-the-art evaluation, through geologic characterization, chemistry, fracturing, and fluid flow related to leaching and solution mining of copper, uranium and precious metals. Biological leaching and leaching of nonmetallics are also covered in this comprehensive study written by experts in the field. Volume 2 contains a bibliography of over 2,000 references, with 300 annotated entries.

431. Vorobjev, B. M., and Deshmukh, R. T. *Advanced Coal Mining.* New York: Asia Publishing House, 1966. 909 p. 2 vols.
Coal mining techniques, especially Russian, are presented for use primarily in underground mines in India. Russian and Indian examples illustrate the techniques.

432. Weiss, Alfred, ed. *Computer Methods for the 80's in the Mineral Industry.* New York: Society of Mining Engineers of AIME, 1979. 975 p.
Mining information systems, exploration, production, planning and management support systems are the focus for this multiauthor work. Computer printout examples are presented, along with flowsheets and many

diagrams illustrating computer applications. Some case studies are included. An extensive, detailed index accompanies the data.

LEGAL AND REGULATORY CONSIDERATIONS

Copies of the relevant mining laws and regulations are necessary in order to evaluate legal requirements and their impact on mineral industries. Interpretations of those laws and regulations by qualified experts assist the manager, engineer and researcher in understanding just what is really required. Both laws and interpretations are included.

433. *Code of Federal Regulations, Part 30, Mineral Resources.* Washington, DC: Government Printing Office, 1985. 3 vols. (Annual)

Mine safety and health regulations, Minerals Management Service requirements, financial aid for exploration, Bureau of Mines grants, and surface coal mining regulations are codified in Code of Federal Regulations (CFR). The compendium of U.S. mining regulations, although other sections of the federal regulations, such as those dealing with the environment, must also be satisfied by mining operations and their personnel.

434. Commerce Clearing House, Inc. *Federal Mine Safety and Health Act of 1977; Law and Explanation.* P.L. 95-164, November 9, 1977. Chicago, IL, 1977. 172 p.

Legal terms and historical background of the act are covered, then sections of the act are explained by subject, such as miner training, safety and health standards. The Conference Report, House & Senate is included for more background on mine safety and health. A detailed index provides access to specific portions of the act and its requirements.

435. Disbrow, David. *Assessment of the Impact of Severance Taxation on Mineral Resource Extraction.* Unpublished M.S. thesis, Colorado School of Mines, 1983. 206 p.

Impact of severance taxes are assessed through modeling a porphyry molybdenum deposit/mine. Severance taxes on a mine's output cause the breakeven cutoff grade of the deposit to increase, thus decreasing mineable reserves. After-tax profits are decreased, thus decreasing other state and federal taxes. Interesting evaluation of an increasingly applied tax and its impacts.

436. Jablonski, Donna M., and Crawford, Mark H., eds. *Federal Coal Leases; Marketing, Management and Financial Profiles.* New York: McGraw-Hill, 1980. 236 p.

Government data covers 558 leases in fourteen states, with lease information on ownership, reserves, quality, mining methods, plans to 2,000, and contract terms taken from the Automated Coal Leasing System maintained by the U.S. Department of the Interior. The five chapters cover financial, marketing/producing leases, marketing/non-producing leases, and production. The index is by lease owner.

437. Jennings, William. *How to Negotiate and Administer a Coal Supply Agreement.* New York: McGraw-Hill, 1981. 522 p.
Practical guide to developing, evaluating, negotiating, administering and litigating coal supply agreements. The focus is western and midwestern, but principles apply throughout the U.S. Sample clauses from coal agreements, such as conditions for rejection of coal or hardships, are planned for people with no experience, but also as a reference for those who are experienced. The author states that there is no optimum contract, so the samples included are only to consider.

438. Joyce, Christopher R. *Final Federal Surface Mining Regulations.* New York: Mine Regulation & Productivity Report and Coal Week, 1980. 100 p.
Eminently readable guide to the Surface Mining Control and Reclamation Act of 1977. Overview of regulations, their scope, purpose and impact. Specific areas of compliance are reviewed and interpreted, such as performance standards, reporting requirements, bonding, permitting, hearings and appeals—all in understandable language.

439. Leshy, John D. *The Mining Law; A Study in Perpetual Motion.* Washington, DC: Resources for the Future, Inc., 1987. 521 p.
A new and eminently readable history of the General Mining Law of 1872 and its influence on the western United States. Changes in the law through interpretation, establishment of wilderness areas, regulations, and continuing reform are all covered in this excellent study. Copious footnotes for each chapter enable tracing the precise legal aspects.

440. Organization of American States, General Secretariat. *Mining and Petroleum Legislation of Latin America and the Caribbean.* Dobbs Ferry, NY: Oceana Publications, Inc., 1979. (Various paging)
Basic mining and petroleum legislation are covered for the twenty-four countries in Latin America and the Caribbean. General principles, exploitation, exploration contracts, taxation, and formal definition of mining companies are presented in this interesting compendium, which can be easily updated periodically because of its loose-leaf format.

441. Post, Alexandra M. *Deepsea Mining and the Law of the Sea.* Publications on Ocean Development, vol 8. The Hague, The Netherlands; Boston; Lancaster, England: Martinus Nijhoff, 1983. 358 p.
Although polymetallic nodules are not profitable now, assuring the free world a supply through political stability is still important. A multilateral law of the sea treaty is proposed, and treaties should be structured to enhance ocean mining in the future.

442. Practising Law Institute. *The Federal Mine Safety and Health Act.* Commercial Law and Practice Course Handbook Series Number 213. New York, 1979. 360 p.
This course book can also serve as a reference manual on the act. Background, procedures for inspectors, judicial proceedings, and investigations, penalties and sanctions are covered. Index is brief.

443. Sammons, Donna. *Coal Industry Taxes; A State-by-State Guide.* New York: McGraw-Hill, 1981. 144 p.
A quick reference to state taxes, with a yes-no checklist, then taxes in order of impact on the coal industry. Each of the twenty-six coal-producing states is ranked from high tax to low. Further information sources and where to go for help are detailed in this practical, easy-to-use guide to a significant aspect of coal mining.

444. Smith, Duane A. *Mining America; The Industry and the Environment, 1800–1980.* Lawrence, KS: University Press of Kansas, 1987. 210 p.
The story of mining, primarily in the western U.S., and mining's attitude toward the environment, from the mining industry's viewpoint. The evolution of environmental attitudes, why industry changed, and societal costs for environmental protection are analyzed in this thoughtful historical study of mining and its impact on the environment. Extensive reference notes and bibliography, plus index.

445. Yasnowsky, P. N. *A Summary of Current and Historical Federal Income Tax Treatment of Mineral Exploration and Development Expenditures.* U.S. Bureau of Mines Information Circular 9011. Washington, DC: Government Printing Office, 1985. 13 p.
Brief summary of trends of mineral taxation from 1951–85.

446. Zuercher, Rick, and Buist, Helena. *Guide to Coal Contracts.* 3d ed. Arlington, VA: Pasha Publications, 1985. 560 p.
Coal contracts are arranged by utility companies which are regulated by Federal Energy Regulatory Commission (FERC). Supplier, source, date, specifications, transportation, rates, producing seams, and mines are covered for 1982–83, with an index of suppliers. Useful for basic data contained in the contracts, as well as examples of functioning contract terms.

RECLAMATION AND ENVIRONMENTAL IMPACT

Reclamation of mine sites, now required, is a political and social issue, as is environmental impact of mining. These studies of mining, reclamation and waste disposal provide insight into current requirements, technology and practices.

447. American Society of Civil Engineers. Committee on Embankment Dams and Slopes of the Geotechnical Engineering Division. *Current Geotechnical Practice in Mine Waste Disposal, Papers.* New York, 1979. 260 p.
Case studies of mine waste disposal in the gold, metal, phosphate, taconite, copper and power plant industries. Current practice is analyzed and recommendations for disposal presented.

448. Bartlett, Robert V. *The Reserve Mining Controversy; Science, Technology, and Environmental Quality.* Bloomington, IN: Indiana University Press, 1980. 293 p.

Public conflicts and environmental questions forced Reserve Mining to stop taconite discharges into Lake Superior. Reserve continued to mine during the controversy. Interesting case study of public concern forcing changes in mining process.

449. Brice, William Charles. *An Analysis Technique for Mineral Resource Planning.* Unpublished Ph.D. dissertation, University of Minnesota, 1981. 774 p.

MINESITE computer program uses twenty-six data sets to evaluate land use and environmental impacts with potential development of a copper-nickel mine in Minnesota. Models for mineral potential, site alternatives, engineering suitability and environmental aspects allow systematic comparison of alternatives. The model is adaptable, with modifications, for other areas.

450. Center, G. W., et al. *Development of Systematic Waste Disposal Plans for Metal and Nonmetal Mines.* U.S. Bureau of Mines Open-File Report 183-82. Pittsburgh, PA: U.S. Bureau of Mines, 1982. 643 p. PB-138065.

Major types of waste disposal are analyzed for an audience of industry, government, and mining consultants. Regulations, technical and economic constraints are considered and alternatives evaluated in this comprehensive study.

451. Down, C. G., and Stocks, J. *Environmental Impact of Mining.* New York: Wiley, 1977. 371 p.

In this British publication, examples of environmental impacts of mining are mostly from the U.K. Emphasis is on surface mining, but some underground impacts are also addressed. Types of pollution are assessed, including air, noise, and water. Mitigating requirements suggested.

452. Doyle, William S. *Deep Coal Mining: Waste Disposal Technology.* Pollution Technology Review No. 28. Park Ridge, NJ: Noyes Data, 1976. 392 p.

Nineteen government reports and seven patents concerning underground coal mines are summarized for acid mine drainage control. Solutions include prevention or control of pollution through sealing, neutralization, underground disposal of waste, and refuse reclamation. Overview of current pollution control technology and practice.

453. Edgar, Thomas F. *Coal Processing and Pollution Control.* Houston, TX: Gulf Publishing, 1983. 579 p.

Coal technology is examined with view toward increasing U.S. coal production. For scientists and engineers, the overall picture of environmental problems and possible solutions leads to a projected increase in U.S. coal use.

454. Honig, Robert A.; Olson, Richard J.; and Mason, William T. *Atlas of Coal/Minerals and Important Resource Problem Areas for Fish and Wildlife in the Coterminous United States.* Washington, DC: U.S. Fish and Wildlife Service, 1981. FWS/OBS-81-06.
Overlay maps showing mineral resources and fish and wildlife resources can be combined to highlight potential problem areas. Clear and effective presentation of conflicting uses and areas where mitigating measures must be carefully planned.

455. Johnson, Allan M.; Laage, Linneas W.; and Butler, Donald James. *The Development of Guidelines for Closing Underground Mines: Executive Summary.* U.S. Bureau of Mines Open-File Report 138(1)-84. Pittsburgh, PA: U.S. Bureau of Mines, 1983. 40 p.
Problems with underground mine closings are presented through case histories. Acid water drainage, subsidence, and inadequately protected mine shafts are all covered, and preplanning recommended to lower environmental impact and mitigation costs.

456. Lee, William W. L. *Decisions in Marine Mining; The Role of Preferences and Tradeoffs.* Cambridge, MA: Ballinger, 1979. 211 p.
Questionnaires were used to assess profit versus risk of sand and gravel mining on the continental shelf, and mathematical calculations used to examine the tradeoff between resources and the environment.

457. U.S. Environmental Protection Agency. *Processes, Procedures, and Methods to Control Pollution from Mining Activities.* Washington, DC: Government Printing Office, 1973. 390 p. EPA-430/9-73-011.
Overview of pollution control techniques in 1973, so alternatives can be identified for future study. The techniques are mostly applicable to coal mining in the eastern U.S.

458. U.S. Office of Technology Assessment. *Western Surface Mine Permitting and Reclamation Summary.* Washington, DC: Government Printing Office, 1986. 54 p.
Because 70 percent of western coal mines include federal coal, OTA examined reclamation technologies and adequacy of baseline data. Policy options are presented, along with cost-benefit data and practical assessments of possibilities and problems in surface coal mine reclamation.

Journals

Journals, along with conferences, provide the most current information on industry trends and technology. General interest and scholarly technical publications published once a year or more frequently are included in this section. Current titles of long-published journals are entered, and former titles cross referenced in the annotation. Journals which have ceased publication and have not been absorbed into another journal are not listed. Current frequency of publication is also the one selected, although frequency may have varied during the life of the publication.

459. *The AIMM Bulletin and Proceedings.* Parkville, Victoria, Australia: Australasian Institute of Mining and Metallurgy, 1898– . (Monthly)
> Divided into two sections: the Bulletin, which features news on people and conferences, and the Proceedings, which contains technical assessments of open pit mining, blasting, and leaching in Australia, New Zealand, the United Kingdom, and the Commonwealth.

460. *AMC Journal.* Washington, DC: American Mining Congress, 1915– . (Monthly)
> The voice of the U.S. mining industry, this management-oriented journal contains very brief articles and timely news items. Previous titles include *American Mining Congress Journal* and *Mining Congress Journal.*

461. *American Metal Market.* New York: AMM, 1882–. (Daily, Monday through Friday)
> This newspaper covers marketing, general business information, and daily prices for scrap and refined products.

462. *Australian Mining.* Chippendale, New South Wales, Australia: Thomson Publications Australia, 1908– . (Monthly)
> Each issue contains news and features on Australian mining and the worldwide mining situation, plus a profile on an Australian mining company.

463. *Australia's Mining Monthly; Incorporating Lodestone's Australian Oil and Gas Journal.* West Leederville, Australia: Mining Monthly House, 1980– . (Monthly)
Nontechnical articles, interviews with mining engineers and managers, and reports on Australian mining companies comprise the *Mining Monthly*.

464. *Black Diamond.* Chicago: Black Diamond Co., 1885– . (Monthly)
Brief news about coal industry, coal production, imports and exports, and management personnel are directed toward executives in coal marketing, production, processing, handling and transportation.

465. *California Mining Journal.* Aptos, CA, 1931– . (Monthly)
Oriented toward scientists, laymen and investors, the short articles and news items combine technical and practical information on California mines and mining.

466. *Canadian Geotechnical Journal.* Ottawa, ON: National Research Council of Canada, 1963– . (Quarterly)
Soil mechanics journal also includes rock mechanics and geology, which are often applicable to mining.

467. *Canadian Mineralogist; Crystallography, Geochemistry, Mineralogy, Petrology, Mineral Deposits.* Toronto, ON: Mineralogical Association of Canada, 1957– . (Quarterly)
Applied mineralogy, which is commercial and practical, is one element of the *Journal of the Mineralogical Association of Canada*. Characteristics of minerals from ore deposits are featured, as well as pure mineralogy. Articles are primarily in English, although French is possible.

468. *Canadian Mining Journal.* Don Mills, ON: Southam Communications Ltd, 1879– . (Monthly)
General information on Canadian mining and mineral industries is presented, along with special features, such as the January summary of geophysical mineral exploration trends and developments.

469. *Chilton's Iron Age Metals Producer.* Radnor, PA: Chilton, 1885– . (Monthly)
Industry-oriented journal covers metals such as aluminum and copper as well as iron and steel prospects, prices, uses and manufacturing. Earlier titles: *Iron Age* and *Chilton's Iron Age*.

470. *CIM Bulletin.* Montreal, Quebec, PQ: Canadian Institute of Mining and Metallurgy, 1898– . (Monthly)
This journal, which is the official bulletin of the Canadian Institute of Mining and Metallurgy, includes technical papers and news briefs on all types of mining and metallurgy in Canada, as well as extensive information on the annual conferences of the Institute. Excellent journal with in-depth coverage especially of engineering aspects of Canadian mining.

471. *Clays and Clay Minerals; Journal of the Clay Minerals Society.* New York: Pergamon, 1968– . (Bimonthly)
Research and technology concerning clays and other fine-grained minerals, their structure, occurrence and applications. Technical and mathematics-based articles on worldwide clay research. Comprehensive subject, author, title index each year. Earlier title: *National Conference on Clays and Clay Minerals Proceedings.*

472. *Coal Outlook.* Arlington, VA: Pasha Publications, 1975– . (Weekly)
This newsletter includes steam coal prices, metallurgical coal prices, coal plants, stock information, legislation, brief current news and business outlook projections. Good coverage of the coal industry.

473. *Coal Week International.* New York: McGraw-Hill, 1975– . (Weekly)
Important and timely newsletter deals with all facets of U.S. coal, from taxes and legislation to technology and markets, plus prices.

474. *Colliery Guardian.* Redhill, England: Fuel and Metallurgical Journals Ltd., 1860– . (Monthly)
Steel and coal information are emphasized in this journal with a practical orientation toward British mining and mineral industry.

475. *Colorado School of Mines (Quarterly).* Golden, CO: Colorado School of Mines Press, 1905– . (Quarterly)
Each issue comprises one lengthy research article on mineral resources, economic geology, mining and engineering.

476. *Earth and Mineral Sciences.* University Park, PA: State University, 1931– . (Quarterly)
News from the university, its faculty and alumni and their activities in the earth sciences are featured, along with general interest articles.

477. *Economic Geology and the Bulletin of the Society of Economic Geologists.* El Paso, TX: Economic Geology Publishing Co., 1906– . (Semiquarterly)
Ore deposits, minerals and their characteristics are covered in scholarly reports, shorter scientific communications, and discussions of previous materials, in this primary journal for economic geologists. Useful to watch trends and discoveries in mineral exploration.

478. *Energy Journal.* Boston: Oelgeschlager, Gunn & Hain, 1980– . (Quarterly)
The Energy Economics Educations Foundation and International Association of Energy Economists publish economic aspects of energy resource development, such as uranium and coal. Both historical and current investigations are featured, and world energy developments are analyzed.

479. *Energy Policy.* Guildford, England: Butterworth, 1973–. (Bimonthly)
This refereed journal covers energy policy and energy planning, with topics such as transportation of western coal in the U.S. or nuclear policies around the world.

480. *Energy Sources; An Interdisciplinary Journal of Science and Technology.* New York: Taylor & Francis, 1973– . (Quarterly)
Trends and solutions to provide worldwide energy, including coal and lignite, in scholarly articles on transportation, gasification and development.

481. *Energy; The International Journal.* Elmsford, NY: Pergamon, 1986– . (Monthly)
Interdisciplinary studies of technologies, resources, demand and policies are geared toward a broad range of readers. The scholarly articles may all treat one topic, such as the November/December 1986 issue, *Newer Coal Technologies; Implications for Energy and Development Policies in Asia and the Pacific.*

482. *Engineering Geology.* Amsterdam, The Netherlands: Elsevier, 1965– . (Quarterly)
Some mining articles are included, generally on mine design, storage, and waste disposal, in this scholarly journal.

483. *Environmental Geochemistry and Health; Incorporating Minerals and the Environment.* Kew, England: Science and Technology Letters, 1979– . (Quarterly)
Under the previous title, *Minerals and the Environment,* contents emphasized the interface between mining and the environment, including reclamation topics. In 1985 the current title was adopted, and emphasis shifted to inorganic influences on public health, but environmental impacts of mining are still covered when applicable to public health.

484. *Fuel Science and Technology International.* New York: Marcel Dekker, 1982– . (6/yr)
Scientific and technical aspects of fuels, including coal and oil shale, as well as heavy oils and biomass, are described by scientists and engineers for technical audiences. In situ production and coal conversion processes included.

485. *Fuel; The Science and Technology of Fuel and Energy.* Guildford, England: Butterworth, 1922– . (Monthly)
Very technical coverage on physical properties, preparation and use of fuels and associated minerals are presented in specialized issues on one type of fuel, such as coal and oil shale.

486. *Gems & Gemology.* Santa Monica, CA: Gemological Institute of America, 1964– . (Quarterly)
　　Jewelry-making and fine gems are the focus of this quarterly, which covers locations, mines, mineral quality and usefulness of gems for jewelry, all in well-referenced articles in the primary journal for the jewelry industry.

487. *Geologie en Mijnbouw; International Journal of the Royal Geological and Mining Society of The Netherlands.* Dordrecht, The Netherlands: Martinus Nijhoff Publishers, 1921– . (Quarterly)
　　Most of the lengthy, scholarly articles are in English, but occasionally in other languages on worldwide mining, mineral deposits, and geological topics, especially on coastal lowlands and adjacent offshore regions.

488. *Geotechnique.* London: Institution of Civil Engineers, 1948– . (Quarterly)
　　Soil mechanics journal with lengthy, scholarly articles, brief technical notes and discussions of earlier articles, primarily by university engineering faculty.

489. *Gluckauf* Essen, West Germany: Verlag Gluckauf, 1865– . (Biweekly)
　　German journal for technology and economics in the mining industry has technical articles now translated into English and bound in the same issue with the German original. German and European mining topics are emphasized.

490. *Gower Federal Service—Mining.* Denver, CO: Rocky Mountain Mineral Law Foundation, 1962– . (Monthly)
　　Loose-leaf format allows continual updating of federal and state laws and regulations regarding mineral claims, plus extensive coverage of legal decisions at all levels.

491. *IM: Industrial Minerals.* Surrey, England: Metal Bulletin Journals Ltd., 1967– . (Monthly)
　　Each issue contains an extensive survey article on one commodity of country's industrial minerals, plus international news and prices of industrial minerals from abrasives to zircon. The basic journal on industrial minerals.

492. *In Situ: Oil, Coal, Shale, Minerals.* New York: Marcel Dekker, 1977– . (Quarterly)
　　Lengthy, detailed technical articles on research and applications indicate the wide variety of in situ mining possibilities, as represented by the title.

493. *Institution of Mining and Metallurgy Transactions, Section A, Mining Industry.* London: IMM, 1892– . (Quarterly)
　　Scholarly, lengthy and well-referenced treatments plus shorter technical notes cover underground, opencast and offshore-mining operations, including finance, planning, management and equipment.

494. *Institution of Mining and Metallurgy Transactions, Section B, Applied Earth Science.* London: IMM, 1892– . (Quarterly)

Theoretical and practical infomation on mineral exploration and mining geology focuses on the United Kingdom with additional international coverage.

495. *Institution of Mining and Metallurgy Transactions, Section C, Mineral Processing and Extractive Metallurgy.* London: IMM, 1892– . Quarterly

Engineering and economic studies of preparation and extraction of ores and minerals focus on practical aspects with lengthy, well-referenced articles and shorter technical notes.

496. *International Geology Review.* Silver Spring, MD: V. H. Winston & Sons, 1958– . (Monthly)

Translations of papers from authoritative Soviet journals plus state-of-the-art reveiws. Economic geology and mineral deposits are often featured.

497. *International Journal for Numerical and Analytical Methods in Geomechanics.* Chichester, England; New York: Wiley, 1977– . (Monthly)

Geomechanics and rock mechanics are an integral part of mining practice, and this scholarly journal covers modeling and other analytical methods applied in geomechanics.

498. *International Journal of Coal Geology.* Amsterdam, The Netherlands: Elsevier, 1980– . (Quarterly)

English-language journal treats coal geology, its petrographic and chemical properties, and coal seams, which are directly relevant to coal exploration, evaluation and mine planning.

499. *International Journal of Mineral Processing.* Amsterdam, The Netherlands: Elsevier, 1974– . (Quarterly)

Lengthy, scholarly articles on processing solid-mineral materials, metallic, nonmetallic and coal.

500. *International Journal of Rock Mechanics and Mining Sciences & Geomechanics Abstracts.* Exeter, England: Pergamon, 1964– . (Bimonthly)

Very technical, often theoretical coverage of rock mechanics is presented in lengthy articles and technical notes of about five pages. Although rock mechanics is emphasized, mining subjects are covered, and the rock mechanics aspects are often applicable to mining.

501. *International Mine Computing.* Sussex, NB, Canada: MM Communications, 1986– . (Bimonthly)

Mining software for personal computers, minicomputers and mainframes is evaluated and samples of programs and printouts are featured.

502. *International Mining.* London: IM, 1984– . (Monthly)
Free to qualified readers in the mining industry, the features, international news and extensive information on equipment are practical and sales-oriented.

503. *Journal of Mines, Metals & Fuels.* Calcutta, India: Books & Journals Private Ltd, 1953– . (Bimonthly)
Indian mining and mineral industries are emphasized, along with relevant international mining news. *Indian Mining Journal* is incorporated within this title.

504. *Journal of the Mine Ventilation Society of South Africa.* Johannesburg, South Africa: MVSSA, 1948– . (Monthly)
Articles in English and Afrikaans cover legislation and new developments concerning mine ventilation in South African gold mines. Often useful for coverage of mine ventilation of deep mines, a technology that South Africa develops highly.

505. *Keystone News Bulletin.* New York: McGraw-Hill, 1974– . (Monthly)
Newsletter reports events in the U.S. coal industry, with brief, timely coverage of industry news, plus current statistics on coal production, consumption and imports. Mines closed are also listed.

506. *Lithology and Mineral Resources.* A Translation of *Litologiya i Poleznye Iskopaemye.* New York: Plenum, 1966– . (Bimonthly)
The original Russian journal on rocks and mineral resources is translated into English within a year of original publication, thus adding to knowledge of world mineral resources.

507. *Marine Mining.* New York: Taylor & Francis, 1977– . (Quarterly)
Undersea mining, not only of manganese nodules, but also of metalliferous soils and deposits, is the focus of this journal, which also covers undersea mineral resources.

508. *Mine and Quarry Mechanisation.* Hamilton, Australia: Magazine Associates. 1960– . (Annual)
Worldwide coverage of all facets of mining and mineral preparation. Many photographs of equipment accompany descriptions of current applications.

509. *Mine & Quarry; Offical Journal of the Minerals Engineering Society.* London: Ashire Publishing, 1924– . (Monthly)
Technical, practical information on coal, aggregate, limestone and other types of mining in the U.K. Extensive personnel and company news. Formerly *Mining & Minerals Engineering, Colliery Engineering, Mine and Quarry Engineering* and *Coal Preparation.*

510. *Mine Safety & Health.* Washington, DC: Government Printing Office, 1975– . (Quarterly)
Published by U.S. Department of Labor, Mine Safety and Health Administration, this quarterly includes training information, accident reports, approved equipment and statistics on health and safety in mining.

511. *Mineral Industry Surveys.* Washington, DC: U.S. Bureau of Mines, 1960– . (Monthly and quarterly)

Various minerals are surveyed, such as gold and silver, with production statistics, new discoveries and coinage in worldwide mineral industries. Number of minerals covered has been expanded through the years.

512 *Mineral Processing and Technology Review.* New York, London: Gordon and Breach, 1983– . (Quarterly)

Reviews primarily on mineral processing and extractive metallurgy emphasize engineering and technology but also consider economics. Scholarly reviews of state-of-the-art technology reported in the literature offer overviews of recent literature and patents.

513. *Mineralium Deposita: International Journal for Geology, Mineralogy and Geochemistry of Mineral Deposits.* Official Bulletin of the Society for Geology Applied to Mineral Deposits. New York: Springer-Verlag, 1966– . (Quarterly)

Scholarly journal is devoted to worldwide mineral deposits and their characteristics, with articles in English, French or German. Lengthy, well-referenced articles often include mining information in addition to geological characteristics of ore deposits.

514. *Minerals & Energy Bulletin; News from the CSIRO Institute of Energy and Earth Resources.* Dickson, Australia: CSIRO, 1980– . (Quarterly)

Brief newsletter describes ongoing research on equipment, exploration and production in the Commonwealth Scientific and Industrial Research Organization (CSIRO), Australia. Contact persons and their telephone numbers are included.

515. *Minerals and Materials; A Bimonthly Survey.* Pittsburgh, PA: U.S. Bureau of Mines, 1975–. (Bimonthly)

World mineral markets are surveyed bimonthly. Plans for new mines are reported, plus monthly production of commodities. Mine closings, postponed plans for mines, and layoffs are also covered.

516. *Minerals & Metallurgical Processing.* Littleton, CO: Society of Mining Engineers of the American Institute of Mining, Metallurgical and Petroleum Engineers, 1984– . (Quarterly)

Technical articles on minerals and related metallurgy published by the primary professional association in the industry.

517. *Mines Magazine.* Golden, CO: Colorado School of Mines Alumni Association, 1910– . (Monthly)

Technical, practical and nonscholarly coverage of mineral projects and processes are included in this monthly which also informs about Colorado School of Mines and its alumni.

518. *Mining Activity Digest.* New York: McGraw-Hill, 1974– . (Monthly)

News about planned mining ventures, personnel changes, mine and plant activities is presented in newsletter format.

519. *The Mining Engineer; Journal of the Institution of Mining Engineers.* Doncaster, England: IME, 1960– . (Monthly)
Technical articles on engineering aspects of British mining emphasize case studies and equipment used in mining in the U.K., as reported by members of the institution. *Transactions of the Institution of Mining Engineers* are incorporated within the journal.

520. *Mining Record.* Denver, CO: Howell Publishing, 1889– . (Weekly)
This weekly newspaper focuses on Colorado and Rocky Mountain mineral industries, including stock prices and data on various companies.

521. *Mining Science & Technology.* Amsterdam, The Netherlands: Elsevier, 1983– . (Quarterly)
Heavily scientific and theoretical journal includes lengthy treatments with tabular data and formulas in worldwide coverage of rock mechanics applications to mining.

522. *Mining Software Review.* Boulder, CO: Rocky Mountain Resource Development, 1982– . (Monthly)
Detailed evaluation of mining software primarily for mainframe computers but including some for personal computers. Examples of results are presented.

523. *Mining Survey.* Johannesburg, South Africa: Chamber of Mines of South Africa, 1982– . (Quarterly)
Annual report type of publication is geared toward the general public, members of the Chamber of Mines, and stockholders in South African mining companies. A particular type of mining, such as coal, may be surveyed as a special feature.

524. *Mining Technology; Journal of the Institution of Mining Electrical and Mining Mechanical Engineers.* Manchester, England: Marylebone Press, 1920– . (Monthly)
Extremely technical information on all phases of mining engineering are geared toward practicing British mining engineers in the fields noted.

525. *Natural Resources Forum.* Published for the United Nations. London: Graham & Trotman, 1976– . (Quarterly)
Major technological and policy issues in international energy, mineral and water resources exploration, development and management are analyzed for policymakers, banking executives, and academics. Country resources and commodity resources are studied.

526. *Northern Miner Magazine.* Toronto, ON: Northern Miner Press, 1986– . (Monthly)
Included in Canada as part of a subscription to *Northern Miner,* this magazine incorporates the former special issues of *Northern Miner* and longer articles on Canadian mining and mining companies.

527. *Pit & Quarry.* Cleveland, OH: Harcourt Brace Jovanovich, 1916– . (Monthly)
Equipment, plants, safety and personnel news in aggregates, sand and gravel, cement, lime, gypsum and nonmetallic minerals industries are reported in this very focused journal.

528. *Power Plant Deliveries.* Washington, DC: National Coal Association, 1973– . (Monthly)
Steam coal prices, cost and quality by region and particular power plant. Sources of coal are listed, even with pit number, to correlate with quality. A numerical picture of the U.S. coal industry.

529. *Quarterly Coal Report.* Washington, DC: Government Printing Office, 1977– . (Quarterly)
Published by the Energy Information Administration of the U.S. Department of Energy, this series reports U. S. coal supply and demand, prices, exports and imports, plus trends of coal supply and demand. Coal mines, manufacturing plants, coal distribution companies, and electric utilities were surveyed to obtain the data.

530. *Quarterly Journal of Engineering Geology.* London: The Geological Society, 1968– . (Quarterly)
This scholarly journal treats all facets of engineering geology, a minor portion of which is applicable to mining.

531. *Resources and Energy.* Amsterdam, The Netherlands: North-Holland, 1978– . (Quarterly)
Scholarly, lengthy and well-referenced treatments of energy resources and their allocation among developed and developing countries. Engineering, social and physical sciences contribute to the interdisciplinary studies focused on economic aspects.

532. *Resources Policy.* Guildford, England: Butterworth, 1975– . (Quarterly)
Resources policy and management, including mineral resources, comprise the scholarly studies, surveys of state industries and their impact, and trade significance of natural resources.

533. *Rock Mechanics and Rock Engineering.* New York: Springer-Verlag, 1963– . (Bimonthly)
Mining applications are included in this scholarly journal oriented to research, with text in English, French and German.

534. *Rock Products.* Chicago: Maclean Hunter, 1897– . (Monthly)
Specific descriptions, costs and specifications are detailed for sand and gravel, aggregate and stone equipment and operations in this thoroughly practical guide for practitioners in the rock industries. The basic journal for sand and gravel mining and quarrying, plus aggregates.

535. *Rocks & Minerals.* Washington, DC: Heldref Publications, 1926– . (Bimonthly)
Geared toward the amateur mineral and fossil collector, specimens are pictured and collecting locations are noted in the nontechnical magazine.

536. *Rocky Mountain Pay Dirt.* Bisbee, AZ: Copper Queen Publishing Co., 1938– . (Monthly)
This informal publication relates business news, history and people news in Colorado, Idaho, Montana, Nevada, Utah and Wyoming. Although the original *Pay Dirt* emphasized copper, the current ones include gold, oil shale, and uranium in addition to copper.

537. *Skillings Mining Review.* Duluth, MN: Skillings, 1912– . (Weekly)
Worldwide brief coverage of mining, with some emphasis on iron and shipping in the Great Lakes region of the U.S. and Canada, also includes mining company stock prices on stock exchanges throughout the world.

538. *South African Mining; Coal, Gold and Base Minerals.* Johannesburg, South Africa: Thomson Publications, 1891– . (Monthly)
Practical articles on management and engineering focus on all types of South African mining. This journal combined *South African Mining and Engineering Journal* and *Coal, Gold and Base Minerals* in 1985.

539. *Southwestern Pay Dirt.* Bisbee, AZ: Copper Queen Publishing Co., 1938– . (Monthly)
This descendant of the original *Pay Dirt* carries news, history, personnel news of all levels of mining employees, and equipment information in an informal manner. Arizona, New Mexico and southern California are the areas, and copper, gold, and other minerals are the commodities in the newsletter format.

540. *Soviet Mining Science.* Translated from *Fiziko-Tekhnicheskie Problemy Razrabotki Poleznykh Iskopaemykh.* New York: Plenum, 1965– . (Bimonthly)
The Russian scholarly publication emphasizing rock mechanics is translated into English within a year of the original publication date.

541. *Sulphur.* London: The British Sulphur Corporation Ltd, 1953– . (Bimonthly)
In-depth information concerning the sulfur and sulfuric acid industry, plants and processes. Includes prices and markets.

542. *Tin International; Mining, Processing, Applications, Equipment, Marketing.* London: Tin Publications, 1928– . (Monthly)
Brief articles on the worldwide tin industry cover statistics on tin production and uses, reviews of equipment used in tin mining and processing, and industry news.

543. *Tunnels & Tunneling.* London: Morgan-Grampian, 1969– . (Monthly)

Particularly useful for equipment information, this journal describes tunneling projects worldwide. Mining transportation by tunnels is one facet covered.

544. *United Mine Workers Journal.* Indianapolis, IN: UMW, 1891– . (Monthly)

The United Mine Workers' point of view stresses employment, negotiations, news, boycotts, and strategies for the union and its members to deal with management.

545. *Uranium; Geology, Exploration, Mining and Milling, and Environmental Aspects.* Amsterdam, The Netherlands: Elsevier, 1982– . (Quarterly)

Lengthy technical treatments focus on uranium exploration and production, its economics, technology and possibilities, with the actual uranium mining and milling a minor portion.

546. *Weekly Coal Production.* Washington, DC: Government Printing Office, 1980– . (Weekly)

Newsletter presents concise statistics on coal production from surface and underground mines in the U.S., by state. Energy Information Administration of the U.S. Department of Energy collects the statistics from the various mining companies.

547. *Western Energy Update.* Denver, CO: Western Interstate Nuclear Board, 1977– . (Bimonthly)

Newsletter surveys energy industry in the fourteen western U.S. states. Nuclear and coal industry notes include federal and state news gleaned from other publications.

548. *World Dredging & Marine Construction.* Irvine, CA: Woodcon, 1965– . (Monthly)

Equipment and new projects, usually with photographs, are described for underwater mining or dredging for commodities such as coal, sand and gravel. General underwater construction is another major focus.

549. *World Mining Equipment.* New York: Technical Publishing, 1975– . (Monthly)

Formed by the merger of *Mining Equipment International, World Mining,* and *World Coal,* this journal focuses on equipment, techniques, and news features concerning all types of mining around the world.

Maps and Atlases

Maps and atlases provide basic overviews of mineral deposits in land and sea areas. Large mines and mining districts may also be shown, as well as infrastructures such as railroads. Sources of data used in compiling the maps are often included, so detailed surveys, maps and other references can be used.

550. African Geological Survey. *Carte Minerale de L'Afrique.* Paris: Organisation des Nations Unies pour L'Education, la Science et la Culture, 1968. Scale 1:10,000,000.
Colored map of mineral deposits, represented according to morphology and substance, such as Au, S. Numbered deposits are listed by name and noted by number on the map. Data in French and English.

551. Asian Research Service. *China's Coal Mining Industry.* Hong Kong, China, 1984. Scale: 1:9,000,000.
Coal mines, their annual production, coal fields, and reserves are displayed on a colored map. Major coal bases included.

552. Australia Bureau of Mineral Resources, Geology and Geophysics. *Metallogenic Map: Australia and Papua New Guinea.* Canberra, Australia, 1972. Scale 1:5,000,000.
Size and form of metal deposits noted on colored geologic map, with metallogenic provinces and chemistry of deposits represented.

553. Derry, Duncan R. *A Concise World Atlas of Geology and Mineral Deposits.* London: Mining Journal Books; New York: Halstead Press, 1980. 110 p.
Ten colored map sheets covering world's land areas show mineral resource distribution and geological controls of genesis and location. Significant mines are included. Distribution of mineral resources is discussed, with emphasis on age, type and resource.

554. Dixon, Colin J. *Atlas of Economic Mineral Deposits.* Ithaca, NY: Cornell University Press, 1979. 143 p.
Forty-eight specific deposits or groups of deposits are described, with cross sections, geologic maps, and profiles. Bibliography included for each deposit. Range of commodities, mostly metalliferous. Schematic world commodity maps also included.

555. McKelvey, V. E. *Preliminary Maps, World Subsea Mineral Resources.* Compiled by V. E. McKelvey and Frank F. H. Wang. Washington, DC: Government Printing Office, 1970. U.S. Geological Survey Miscellaneous Geologic Investigations Map I-632. 4 sheets, scale 1:60,000,000 at Equator.

Phosphorite, manganese-oxide nodules and metal-bearing muds are delineated in color. Subsea underground mines and coastal placer deposits on one sheet. Useful compilation of worldwide data on subsea resources.

556. *Mineral Distribution Map of Asia.* 2d ed. New York: United Nations Economic and Social Commission for Asia and the Pacific, 1979. 3 sheets, scale: 1:5,000,000.

Compiled from data from the Geological Surveys. Mineral deposits shown by type and substance, and surrounding geologic formations are given in color on colored geologic maps.

557. *Oxford Economic Atlas of the World.* 4th ed. London: Oxford University Press, 1972.

Minerals and metals, energy resources maps are included along with more general maps. Exports and imports listed, along with major producers, and amount of production.

558. Whitmore, Duncan Richard Elmer. *Mineral Deposits of Canada.* Ottawa: Geological Survey of Canada, 1969. Scale 1:5,000,000.

Colored map of mineral deposits, noted by relative size, geologic setting, such as sedimentary, and geologic type, such as placers. Producers of over 36,000 tons are listed and numbered by province.

Associations and Research Centers

In mining and mineral industries many of the research centers provide information to update and supplement published sources. U.S. government agencies are often very useful, especially for data on public lands; state geological surveys and mining bureaus are similarly helpful within their jurisdictions. Specific listings for federal agencies are given, and sources to find state agencies included. Sources for international centers are also included.

In addition, the professional societies and associations can be very helpful, sometimes with statistics, sometimes in noting industry trends and attitudes. National associations have been listed; state and international listings are indicated.

Associations and research centers are resources for quick and current information, with experts in their particular area of interest. Telephone numbers for the headquarters are included in this section.

American Mining Congress. 1920 N St. N.W., Suite 300, Washington, DC 20036 (202) 861-2800.
Broad membership of U.S. coal, metals and mineral producers, plus mining equipment manufacturers and engineering firms. Useful for information on mining equipment, plus industry positions on mining issues.

National Coal Association. 1130 17th St. N.W., Washington, DC 20036 (202) 463-2625.
Association of U.S. coal producers, sellers and transporters gives industry's reaction to coal legislation and policy. Useful for statistical data on production, also for coal traffic and transportation information.

National Sand and Gravel Association. 900 Spring St., Silver Spring, MD 20910 (303) 587-1400.
Sand and gravel producers and distributors. Useful for technical information on equipment and methods for sand and gravel mining. The association also compiles data on computer hardware and software in the sand and gravel industry.

Society of Mining Engineers of AIME. Caller No. D, Littleton, CO 80127 (303) 973-9550.
This member society of The American Institute of Mining, Metallurgical and Petroleum Engineers includes mining engineers in all minerals ex-

cept petroleum. Useful for statistics and updated directory of its members and their addresses beyond that published in *Mining Engineering* each July.

Solution Mining Research Institute. 812 Muriel St., Woodstock, IL 60098 (815) 338-8579.
Small group of salt and chemical companies and those companies that serve the solution mining and underground storage industries. Very useful for locating elusive materials on solution mining.

U.S. Bureau of Land Management (BLM). Department of the Interior, Washington, DC 20240 (202) 343-9435.
Because the Bureau of Land Management (BLM) manages the 342 million acres of public lands in the U.S., mineral management of those lands is included along with such other resources as timber, wildlife and vegetation. BLM also manages the subsurface resources, both mineral and energy, of other areas where the U.S. government owns the mineral rights. Mining claims records for public lands are maintained by, and the information is available for research at, field offices primarily in the western states.

U.S. Bureau of Mines. Department of the Interior, 2401 E St., N.W., Washington, DC 20241 (202) 634-1004.
Because the Bureau is heavily involved in mining research, it is an excellent source for information on current and possible future mining technology. Nonfuel mineral supplies are also tracked and evaluated. Probably the most useful information to be obtained from the bureau is the myriad of statistical data the bureau collects, and current information can be obtained by telephone. Commodity experts can provide insight as well as data, for example, J.M. Lucas, Gold, (202) 634-1070, and R.G. Reese, Silver, (202) 634-1071. International information can be obtained from the Division of Foreign Data at the Washington address. The Bureau of Mines maintains several libraries, and the librarians are very helpful in locating references and bureau experts in mineral industries.

U.S. Department of Energy. 1000 Independence Ave., Washington, DC 20585 (202) 252-5000.
Responsible for research and development of the nation's energy resources, including coal and nuclear, in high-risk, high-return technology to be applied by the private sector. For information on fossil energy, telephone (301) 353-2617; for nuclear energy information, telephone (202) 252-6452.

U.S. Energy Information Administration (EIA). Washington, DC (202) 252-2363.
Through its many publications on energy production and consumption, EIA provides extensive current statistical data. More current information can be obtained to update publications. Very useful also for energy trends and general information on energy.

U.S. Geological Survey. Department of the Interior, 119 National Center, Reston, VA 22092 (703) 648-4460.
Primarily useful for basic data on mineral resources through its geologists working in the field, and for extensive holdings of geologic information in its network of libraries headed by the one in Reston, VA.

The United States Government Manual. Washington, DC: Government Printing Office. (Annual)
This annual directory of U.S. government agencies lists headquarters, regional, area and branch offices along with addresses and telephone numbers. This is an excellent place to start when seeking information not available in printed sources. Several calls or transferred calls will probably be necessary before locating the person who has the answers.

U.S. Mine Safety and Health Administration (MSHA). Department of Labor, Room 601, 4015 Wilson Blvd., Arlington, VA 22203 (703) 235-1452.
Useful for data on current safety practices in all types of mining, including metal, nonmetal and coal. Because MSHA is involved in health and safety research, new information on health procedures can be obtained. In addition, safety materials are available in its regional libraries.

U.S. Minerals Management Service. 12203 Sunrise Valley Dr., Washington, DC 22091 (202) 343-3983.
A relatively recent service established in 1981 to evaluate, classify and lease offshore minerals. Through such resources as Presto computer model, generates estimates of undiscovered resources.

U.S. Office of Surface Mining Reclamation and Enforcement (OSMRE). Department of the Interior, 18th & C Streets, N.W., Washington, DC 20240 (202) 343-4719.
OSMRE deals with coal mining reclamation, either through checking to see that states are properly enforcing reclamation regulations or directly in those states that have not assumed responsibility for coal mine reclamation. A call to Washington will determine if the state regulates or the federal government and relevant regulations can be obtained.

Worldwide Directory of Mine Bureaus and Geological Surveys. Section 4, *E&MJ International Directory of Mining.* (Annual)
About fifteen pages of listings for U.S. Bureau of Mines, including field offices, U.S. Geological Survey, U.S. Mine Safety and Health Administration, including district offices. State mine bureaus and geological surveys also listed, plus international mine bureaus arranged by country. Mining associations, both U.S. and international, are also included.

Yearbook of International Organizations. Brussels, Belgium: Union of International Associations, 1948– . 3 vols. (Annual)
International organizations on mineral industries are listed, first under general minerals, ores, and then by precious stones, energy. Valuable for locating industrial groups interested in minerals. Many professional organizations also included. Subject keyword index, in English and French. Broad coverage includes many organizations relevant to mining and minerals.

Sources

The following list is composed of publishers and organizations that compile information relevant to the mining industry. Many of these are outside regular publication channels, so their addresses are included here.

A. A. Balkema
Boston, MA

Academic Press
Orlando, FL 32887
(305) 345-4100

Akademiai Kiado
Postafiok 24
H-1363
Budapest, Hungary

Almquist & Wicksell
Box 45150
S-10430
Stockholm, Sweden

Aluminum Association
818 Connecticut Ave., N.W.
Washington, DC 20006

American Bureau of Metal Statistics
400 Plaza Dr.
PO Box 1405
Secaucus, NJ 07094-0405

American Geological Institute
4220 King St.
Alexandria, VA 22302

American Iron and Steel Institute
1000 16th St., N.W.
Washington, DC 20036

American Iron Ore Association
1501 Euclid Ave., 514 Bulkley Bldg.
Cleveland, OH 44115

American Metal Market
7 E. 12th St.
New York, NY 10003

American Mining Congress
1920 N St., N.W.
Washington, DC 20036

American Society for Engineering
   Education
11 Dupont Circle, Suite 200
Washington, DC 20036

American Society for Metals
Metals Park, OH 44073

The American Society of Civil
   Engineers
345 E. 47th St.
New York, NY 10017

Amerind Publishing Co. Pvt. Ltd
66 Janpath
New Delhi 110001 India

Arlington House
One Park Ave.
New York 10016

Ashire Publishing Ltd.
42 Grays Inn Road
London WC1X 8LR England

Asia Publishing House
Apts Books Inc.
141 E. 44th St., Suite 511
New York, NY 10017
(212) 697-0887

Association of Exploration
  Geochemists
PO Box 523
Rexdale, Ontario M9W 5L4 Canada

Australasian Institute of Mining
  and Metallurgy
PO Box 122
Parkville, Victoria 3052 Australia

Ballinger Publishing Co.
54 Church St., Harvard Square
Cambridge, MA 02138
(617) 492-0670

Black Diamond Co.
343 S. Dearborn St.
Chicago, Il 60604

Blackwell Scientific Publications
52 Beacon St.
Boston, MA 02108

Books & Journals Private Ltd.
6/2 Madan St.
Calcutta, India 700072

The British Sulphur Corp. Ltd.
Parnell House, 25 Wilton Rd.
London SW1V 1NH England

Broker's Guide Publications
921 S.W. Washington St., Suite 720
Portland, OR 97205
(503) 295-0131

BRS, BRS After Dark,
  BRS/BRKTHRU
BRS Information Technologies
1200 Route 7
Latham, NY 12110
(518) 783-1161

Bureau of National Affairs
1231 25th St., N.W.
Washington, DC 20037

Butterworth Publishers
80 Montvale Ave.
Stoneham, MA 02180

Cadmium Association
34 Berkeley Square
London W1X 6AJ England

California Mining Journal
PO Box 2260
Aptos, CA 95001

Cambridge Scientific Abstracts
5161 River Rd.
Bethesda, MD 20816

Canadian Institute of Mining and
  Metallurgy
400—1130 Sherbrooke St.
Montreal, Quebec H3A 2M8
  Canada

CAN/OLE (Canadian Online
  Enquiry Service)
National Research Council Canada
Ottawa, Ontario K1A 0S2 Canada
(613) 993-1210

CANMET
500 Booth St.
Ottawa, Ontario, K1A 0G1 Canada

CBI Publishing Co.
51 Sleeper St.
Boston, MA 02210

Chamber of Mines of South Africa
Public Affairs Department
PO Box 809
Johannesburg, South Africa 2000

Chapman and Hall
733 Third Ave.
New York, NY 10017

Chase Econometrics
150 Monument Rd.
Bala Cynwyd, PA 19004
(215) 667-6000

Chemical Abstracts Service
2540 Olentangy River Rd.
Columbus, OH 43210
(614) 421-3600

Chemicon Publishing Lt.
Pernell House, 25 Wilton Rd.
London SW1X 1 NH England

Chilton Co.
Chilton Way
Radnor, PA 19089

Cii House
31 Theobalds Rd.
London WC1 England

Clay Minerals Society
P.O. Box 595
Clarkson, NY 14430

Colorado School of Mines Press
Golden, CO 80401
(303) 273-3000

Commerce Clearing House, Inc.
4025 W. Peterson Ave.
Chicago, IL 60646

Congressional Quarterly
1414 22nd St., N.W.
Washington, DC 20037

Consolidated Gold Fields PLC
31 Charles II St.
St. James's Square
London SW1Y 4AG England

Consultants Bureau, New York and
London
233 Spring St.
New York, NY 10013

Copper Queen Publishing Co.
PO Drawer 48
Bisbee, AZ 85603

Council on Economics and National
Security (CENS)
1730 Rhode Island Ave. N.W.
Washington, DC 20036

Crossroads Press
Epstein Bldg, Brandeis University
Waltham, MA 02154

CSIRO Institute of Energy and
Earth Resources
PO Box 225
Dickson ACT 2602 Australia

CSM Alumni Association
Colorado School of Mines
Chauvenet Hall
Golden, CO 80401

DATA-STAR
D-S Marketing Ltd.
Plaza Suite, 114 Jermyn St.
London SW1Y 6HJ England
44 (1) 930-5503

DIALOG Information Services, Inc.
3460 Hillview Ave.
Palo Alto, CA 94304
(415) 858-3785

D. Reidel Publishing Co.
160 Old Derby St.
Hingham, MA 02043

Economic Geology Publishing Co.
202 Quinn Hall
University of Texas at El Paso
El Paso, TX 79968

Economic Information & Agency
342 Hennessy Rd.
Hong Kong

EIC/Intelligence
48 W. 38th St.
New York, NY 10018

E. I. DuPont de Nemours & Co.
Wilmington, DE 19898

Elsevier Science
PO Box 330
1000 AH Amsterdam
The Netherlands

Elsevier Science Publishing Co.
Journal Information Center
52 Vanderbilt Ave.
New York, NY 10017

Engineering Information, Inc.
345 E. 47th St.
New York, NY 10017

Environmental Policy Institute
317 Pennsylvania Ave., S.E.
Washington, DC 20003

ESA—IRS
C.P. 64 Via Galileo Galilei
00044 Frascati, Italy
39 (6) 940 11

E. Schweizerbart'sche
Verlagsbuchhandlung
Johannesstrasse 3A
7000 Stuttgart 1, Germany

Fairchild Publications
7 East 12th St.
New York 10003

Financial Post Corporation Service
PO Box 100, Station A
Toronto, Ontario M5W 1A7
Canada

Financial Times Ltd.
Bracken House, 10 Cannon St.
London EC4P 4BY England

Fuel & Metallurgical Journals
  Limited
Queensway House, 2 Queensway
Redhill, Surrey RH1 1QS England

Gemological Institute of America
1660 Stewart St.
Santa Monica, CA 90404

General Electric Information
  Services Co.
401 N. Washingtron
Rockville, MD 20850

The Geological Society
Burlington House
Picadilly, London W1V 0JU
  England

Geosystems
PO Box 1024
Westminster, London SW1P 2JL
  England
44 (1) 222-7305

Gold '86
55 University Ave., Suite 1700
Toronto, Ontario M5J 2H7
Canada

Gordon and Breach Science
  Publishers
OPA Ltd.
61 Grays Inn Rd.
London WC1X 8TL England

Gower Publishing Ltd.
Gower House, Croft Rd.
Aldershot, Hants., England

Graham & Trotman Ltd.
Sterling House, 66 Wilton Rd.
London SW1V 1DE England

Gulf Publishing Co.
P.O. Box 2608
Houston, TX 77001
(713) 529-4301

Harcourt Brace Jovanovich
7500 Old Oak Blvd.
Cleveland, OH 44130

Hart Publications, Inc.
PO Box 1917
Denver, CO 80201
(303) 837-1917

Health and Safety Executive
Library, Safety Engineering
  Laboratory
Red Hill of Broad Lane
Sheffield S3 7HQ England

Heldref Publications
4000 Albemarle St., N.W.
Washington, DC 20016

HMSO (Her Majesty's Stationery
  Office) Publications Centre
PO Box 276
London SW8 5DT England

Holt, Rinehart and Winston
521 Fifth Ave.
New York, NY 10175

Howell Publishing Co.
311 Steele St., Suite 208
Denver, CO 80206
(303) 355-5202

H. W. Wilson Co.
950 University Ave.
Bronx, NY 10452

IEA Coal Research (International
  Energy Agency)
14/15 Lower Grosvenor Place
London SW1W 0EX England

IIT Research Institute
P.O. Box 4963
Chicago, IL 60680

Indiana University Press
10th & Morton Sts.
Bloomington, IN 74403
(812) 335-8287

Information Plus
Data Resources, Inc.
1750 K St., N.W., Suite 1060
Washington, DC 20006
(202) 862-3720

Information Resources Press
2100 M St., N.W.
Washington, DC 20037

Institution of Civil Engineers
Thomas Telford Ltd.
PO Box 101, 26-34 Old St.
London EC1P 1JH England

Institution of Mining and
  Metallurgy
44 Portland Place
London W1N 4BR England

Institution of Mining Engineers
Danum House, South Parade
Doncaster DN1 2DY England

Intermediate Technology
  Publications Ltd.
9 King St.
London WC2E 8HN England

International Atomic Energy
  Agency
Wagramer St. 5, Box 100 A-1400
Vienna, Austria

International Energy Agency
London, England

I. P. Sharp Associates
1200 First Federal Plaza
Rochester, NY

ISI (Institute for Scientific
  Information)
3501 Market St.
Philadelphia, PA 19104

J. Dick & Co.
Subsidiary of International
  Thomson Organization
12 Lunar Dr., Drawer AB
Woodbridge, CT 06525
(203) 397-2600

Knowledge Index
DIALOG Information Services, Inc.
3460 Hillview Ave.
Palo Alto, CA 94304
(415) 858-3796

The Law Book Co. Limited
44-50 Waterloo Rd., North Ryde,
  New South Wales, Australia
389-393 Lonsdale St., Melbourne,
  Victoria, Australia
6 Sherwood Court, Perth, Western
  Australia

Lead Development Association
34 Berkeley Square
London W1X 6AJ England

Lexington Books
125 Spring St.
Lexington, MA 02173
(617) 862-6650

Longman Group Limited
Westgate House, The High
Harlow, Essex CM20 1NE England

Maclean Hunter Publishing Co.
300 W. Adams St.
Chicago, IL 60606

Macmillan Publishers
Little Essex St.
London WC2R 3LF4 England

Marcel Dekker
270 Madison Ave.
New York, NY 10016

Marquis Who's Who
200 East Ohio St.
Chicago, IL 60611

Martin Consultants, Inc.
PO Box 1076
Golden, CO 80402

Martin Marietta Data Systems
6303 Ivy Lane
Greenbelt, MD 20770
(301) 982-6500

Martinus Nijhoff Publishers
PO Box 322
3300 AH Dordrecht
The Netherlands

Marylebone Press Ltd.
18 Lloyd St.
Manchester M2 5WA England

McClelland and Stewart Limited
25 Hollinger Rd.
Toronto, Ontario M4B 3G2 Canada

McGraw-Hill
1221 Avenue of the Americas
New York, NY 10020

Mead Data Central, Inc.
PO Box 933
Dayton, OH 45401
(513) 865-6800

Metal Bulletin Books Ltd.
Park House, Park Terrace
Worcester Park, Surrey KT4 7HY
England

Metal Bulletin Journals Ltd.
708 3rd Ave.
New York, NY 10017
See also Metal Bulletin Books Ltd.

Metals Economics Group, Ltd.
1722 Fourteenth St.
Boulder, CO 80302
(303) 442-7501

Miller-Freeman Publications, Inc.
500 Howard St.
San Francisco, CA 94105

Miller Freeman Publications Ltd.
69-71 Wembley Rd.
Wembley, Middlesex HA9 8BL
England

Mine Ventilation Society of South
Africa
PO Box 61019, 2 Hollard St.
Marshalltown, Johannesburg 2107
South Africa

Mineral Land Publications
PO Box 1186
Boise, ID 83701

Mineralogical Association of
Canada
Royal Ontario Museum
Toronto, Ontario M5S 2C6 Canada

Mineralogical Society
41 Queen's Gate
London SW7 4HR England

The Mining and Metallurgical
Society of America
230 Park Ave., Suite 1354
New York, NY 10164

The Mining Journal Limited
60 Worship St.
London EC2A 2HD England

Mining Monthly House
PO Box 78
Leederville 6007, Perth, Western
Australia

MM Communications
63 Broad St., Box 1698
Sussex, New Brunswick E0E 1P0
Canada

Montana Bureau of Mines and
Geology
Montana College of Mineral Science
and Technology
Butte, MT 59701

Morgan-Grampian Ltd.
30 Calderwood St.
London SE18 6QH England

National Coal Association
1130 Seventeenth St., N.W.
Washington, DC 20036-4677
(202) 463-2640

National Geophysical Data Center,
NOAA
E/GC Code 291
325 Broadway
Boulder, CO 80303

National Research Council of
Canada
Ottawa, Ontario K1A 0R6 Canada

National Technical Information
Service
5285 Port Royal Rd.
Springfield, VA 22161
(202) 724-3509

Neal Schuman
23 Cornelia St.
New York, NY 10014
(212) 620-5990

New Mexico Institute of Mines and
Technology
Socorro, NM 87801

Nichols Publishing
PO Box 96
New York, NY 10024

North-Holland
Elsevier Science Publishers BV
Postbus 2400
1000CK Amsterdam
The Netherlands

Northern Miner Press
7 Labatt Ave.
Toronto M5A 3P2, Ontario Canada

Northwest Mining Association
633 Peyton Bldg.
Spokane, WA 99201

Noyes Data Publications
Mill Rd.
Park Ridge, NJ 07656

Oceana Publications, Inc.
75 Main St.
Dobbs Ferry, NY 10522
(914) 693-1733

Oelgeschlager, Gunn & Hain, Inc.
131 Clarendon St.
Boston, MA 02116
(617) 437-9620

Ontario Ministry of Natural
    Resources
Toronto, Ontario

ORBIT Information Technologies
    Corporation
1340 Old Chain Bridge Rd.
McLean, VA 22101
(703) 442-0900

Oxford University Press
200 Madison Ave.
New York, NY 10016
(212) 564-6680

Padley & Venables Ltd.
Callywhite Lane
Dronfield, Sheffield S18 6XT
    England

Pasha Publications
1401 Wilson Blvd., Suite 910
Arlington, VA 22209

Pergamon InfoLine Inc.
1340 Old Chain Bridge Rd.
McLean, VA 22101
(703) 442-0900

Pergamon Journals Inc.
Maxwell House, Fairview Park
Elmsford, NY 10523

Pergamon Journals Ltd.
Hennock Rd., Marsh Barton
Exeter, Devon EX2 8NE England

Pergamon Press
Maxwell House, Fairview Park
Elmsford, NY 10523
(914) 592-7700

Plenum Publishing Corp.
233 Spring St.
New York, NY 10013

Practising Law Institute
810 Seventh Ave.
New York, NY 10019

Praeger Publications
521 Fifth Ave.
New York, NY 10017

Predicasts
200 University Circle, Research
    Center, 11001 Cedar Ave.
Cleveland, OH 44106

Public Affairs Information Service
11 W. 40th St.
New York, NY 10018

QL Systems Ltd.
112 Kent St., Suite 205, Tower B
Ottawa, Ontario K1P 5P2 Canada
(613) 238-3499

Rocky Mountain Mineral Law
    Foundation
Porter Administration Bldg.
7039 E. 18th Ave.
Denver, CO 80220

Rocky Mountain Resource
    Development
PO Box 1234
Golden, CO 80401

STN International
c/o Chemical Abstracts Service
2540 Olentangy River Rd.
PO Box 3012
Columbus, OH 43210
(614) 421-3600

Science and Technology Letters
12 Clarence Rd.
Kew, Surrey England

Science Reviews, Inc.
707 Foulk Rd., Suite 102
Wilmington, DE 19802

Silver Institute
1001 Connecticut Ave., N.W., Suite
  1138
Washington, DC 20036

Skillings Mining Review
First Bank Place, Suite 728
Duluth, MN 55802

Society of Mining Engineers of The
  American Institute of Mining,
Metallurgical and Petroleum
  Engineers
8307 Shaffer Parkway
Littleton, CO 80127

Society of Mining Geologists of
  Japan
Nihon Kogyo Kaikan Bldg.
Ginza 8-5-4
Chuo-Ku, Tokyo 104 Japan

Southam Business Publications
1450 Don Mills Rd.
Don Mills, Ontario M3B 2X7
  Canada

Spon (Methuen)
29 W. 35th St.
New York 10001
(212) 244-3336

Taylor & Francis
3 E. 44th St.
New York, NY 10017

TECH DATA
Information Handling Services
Department 438, 15 Inverness Way
  East
PO Box 1154
Englewood, CO 80150
(303) 790-0600

Technical Publishing
Dun & Bradstreet
875 Third Ave.
New York, NY 10022

Telesystemes-Questel
83-85 Boulevard Vincent Auriol
75013 Paris, France
33 (1) 45 82 64 64

Terraspace Inc.
304 N. Stonestreet Ave.
Rockville, MD 20850

Thomson Publications Australia
47 Chippen St.. Box 65
Chippendale, New South Wales
  20008

Thomson Publications SA (Pty)
  Ltd.
PO Box 8308
Johannesburg, South Africa 2000

Tin Publications Ltd.
222-225 The Strand
London WC2R 1BA England

Trans Tech
16 Bear Skin Neck
Rockport, MA 01966
(617) 546-6426
Clausthal-Zellerfeld, Federal
  Republic of Germany

United Mine Workers
2457 E. Washington
Indianapolis, IN 46201

U.S. Bureau of Mines
4800 Forbes Ave.
Pittsburgh, PA 15213

U.S. Department of Energy, Energy
  Information Administration
Forrestal Building E1-22
Washington, DC 20585

U.S. Government Printing Office
Washington, DC 20402

University of Arizona Press
1615 E. Speedway
Tucson, AZ 85719
(602) 621-1441

University of Texas Press
Box 7819
Austin, TX 78713
(512) 471-4032

University of Tulsa, Information
  Services Division
600 S. College
Tulsa, OK 74104

University Press of America
PO Box 19101
Washington, DC 20036

Van Nostrand Reinhold
115 Fifth Ave.
New York, NY 10003

VCH Publishers
303 N.W. 12th Ave.
Deerfield, FL 33442

Verlag Gluckauf GmbH
PO Box 103945
D-4300 Essen 1, Federal Republic
of Germany

West Virginia University
Department of Mining Engineering
Morgantown, WV 26506

Western Interstate Nuclear Board
6500 Stapleton Plaza, 3333 Quebec
St.
Denver, CO 80207

Western Mine Engineering
PO Box 9008
Spokane, WA 99209

Westview Press
5500 Central Ave.
Boulder, CO 80301

Wiley
605 Third Ave.
New York, NY 10158

WILSONLINE
The H.W. Wilson Co.
950 University Ave.
Bronx, NY 10452

Woodcon
Symcom Publishing
Box 17479
Irvine, CA 92713

Zinc Development Association
34 Berkeley Square
London W1X 6 AJ England

Author Index

Numbers refer to citation numbers, not page numbers.

Title Index

Numbers in italic refer to page numbers. All other numbers refer to citation numbers.

Subject Index

Numbers in italic refer to page numbers. All other numbers refer to citation numbers.